NATURE TRANSFORMED

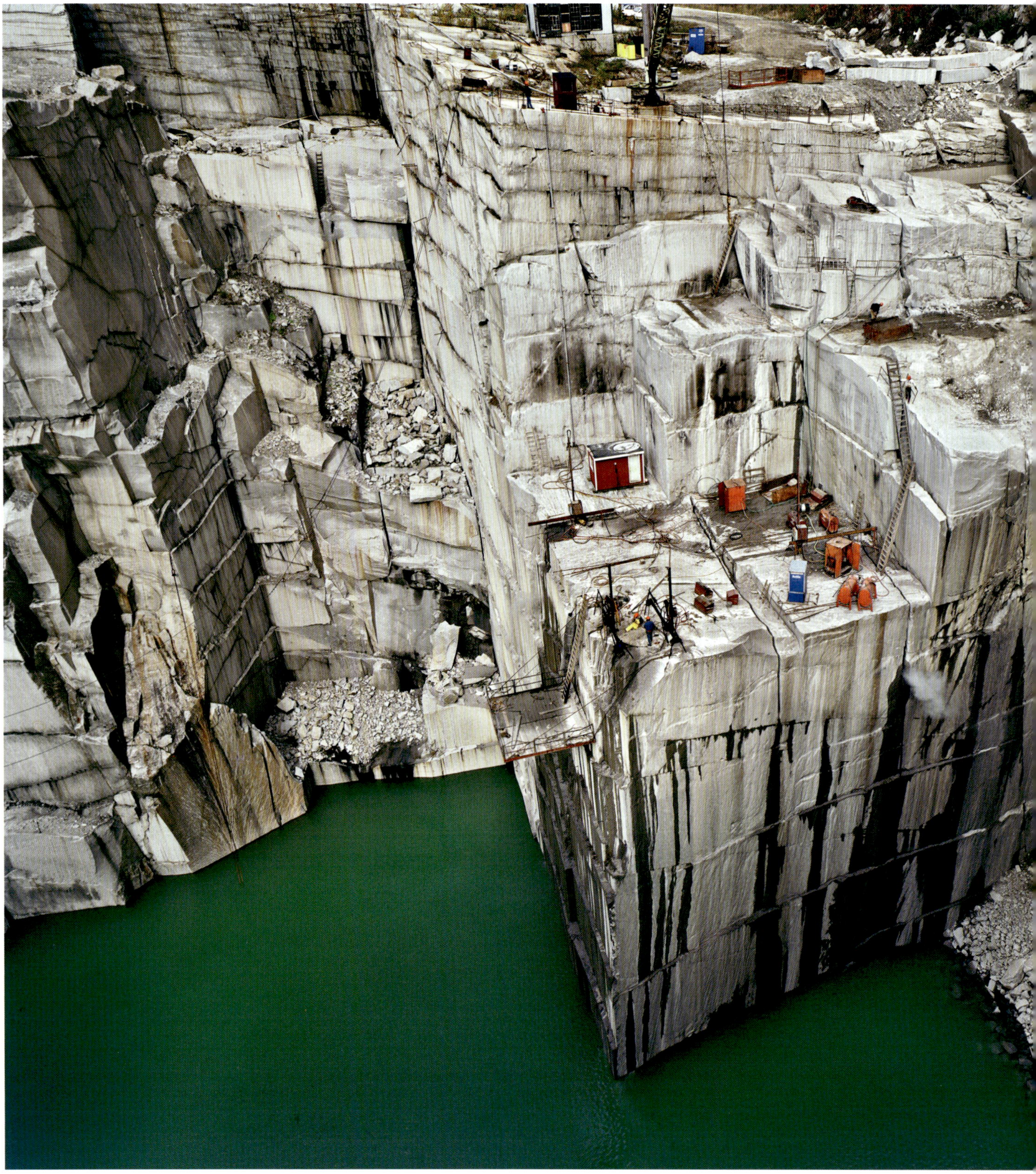

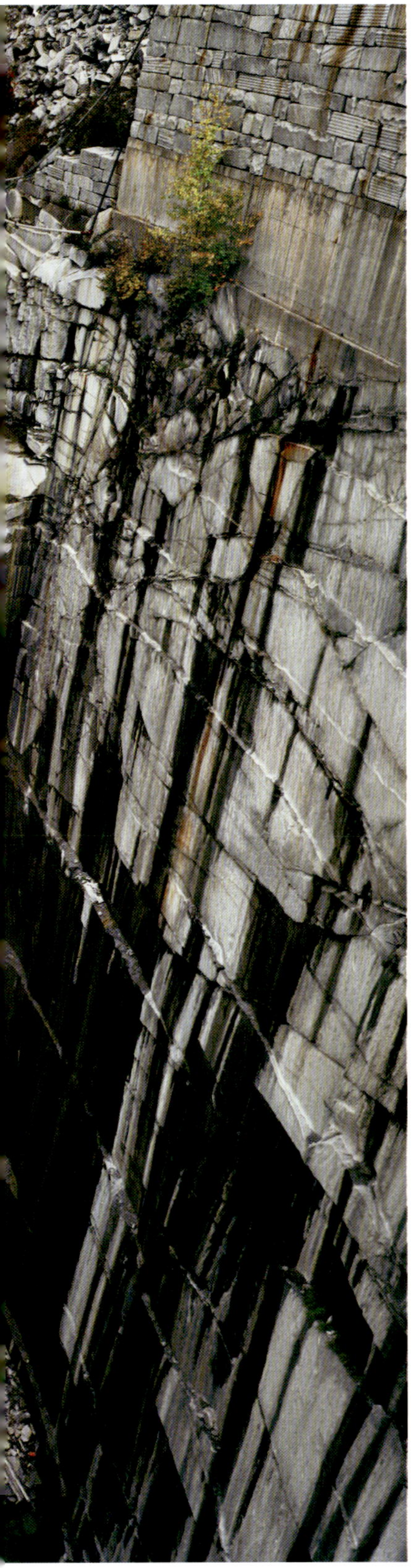

NATURE TRANSFORMED

Edward Burtynsky's Vermont Quarry Photographs in Context

Edited by
Juliette Bianco and Pieter Broucke

With contributions by
Ilaria Brancoli Busdraghi
Kirsten Hoving
Gary D. Johnson

Hood Museum of Art, Dartmouth College
Hanover, New Hampshire

Distributed by University Press of New England
Hanover and London

Published by
Hood Museum of Art, Dartmouth College
Hanover, N.H. 03755
www.hoodmuseum.dartmouth.edu

Distributed by
University Press of New England
One Court Street, Lebanon, N.H. 03766
www.upne.com

The exhibition *Nature Transformed: Edward Burtynsky's Vermont Quarry Photographs in Context* was organized by the Hood Museum of Art, Dartmouth College. The exhibition and catalogue were generously supported by Raphael and Jane Bernstein / Parnassus Foundation, Laurie Jean Weil D.V.M. in honor of her parents, Jean and Bucks Weil, Dartmouth Class of 1935, the Marie-Louise and Samuel R. Rosenthal Fund, and the Ray Winfield Smith 1918 Fund.

Exhibition Schedule:

Hood Museum of Art, Dartmouth College
April 21–August 19, 2012

Middlebury College Museum of Art
February 8–April 21, 2013

Edited by Nils Nadeau
Designed by Glenn Suokko
Printed by Capital Offset Company

Frontispiece:

Edward Burtynsky, *Rock of Ages #7, Active Granite Section, E. L. Smith Quarry, Barre, Vermont, 1991,* digital chromogenic color print. Courtesy of the artist.

Unless otherwise noted, all historic photographs are courtesy of the Aldrich Public Library, Barre, Vermont. All photographs by Edward Burtynsky are courtesy Howard Greenberg & Bryce Wolkowitz, New York / Nicholas Metivier, Toronto. Photograph on p. viii by Juliette Bianco.

ISBN 978-0-944722-43-5

Printed in the United States of America

Library of Congress Cataloguing-in-Publication Data

Burtynsky, Edward, 1955–
Nature transformed : Edward Burtynsky's Vermont quarry photographs in context / edited by Juliette Bianco and Pieter Broucke ; with contributions by Ilaria Brancoli Busdraghi, Kirsten Hoving, Gary Johnson.
p. cm.
Includes bibliographical references.
ISBN 978-0-944722-43-5
1. Landscape photography—Vermont—Exhibitions. 2. Burtynsky, Edward, 1955– —Exhibitions. 3. Quarries and quarrying—Vermont—Pictorial works—Exhibitions. 4. Stone industry and trade—Vermont—Pictorial works—Exhibitions. I. Bianco, Juliette M. II. Broucke, Pieter B. F. J. III. Title.
TR660.B87 2012
779.092—dc23

2011052481

Contents

Preface and Acknowledgments vii
Michael R. Taylor

1 Introduction: Platonic Voids—Ecological Scars 1
Edward Burtynsky and the Quarries of Vermont
Pieter Broucke

2 Digging Deep 9
Edward Burtynsky's Vermont Quarry Photographs
Kirsten Hoving

I Plates: Vermont Marble Quarries 18

3 Proctor and Barre 25
The Marble and Granite Quarries and the Italian Stoneworkers in Vermont, 1880–1915
Ilaria Brancoli Busdraghi

II Plates: Carrara Marble Quarries 36

4 Appalachian Legacy 41
The Uncovered Record of Mountain Building in New England
Gary D. Johnson

III Plates: Vermont Granite Quarries 49

5 An Interview with Edward Burtynsky 61
Juliette Bianco

Selected Bibliography 67

Exhibition Checklist 69

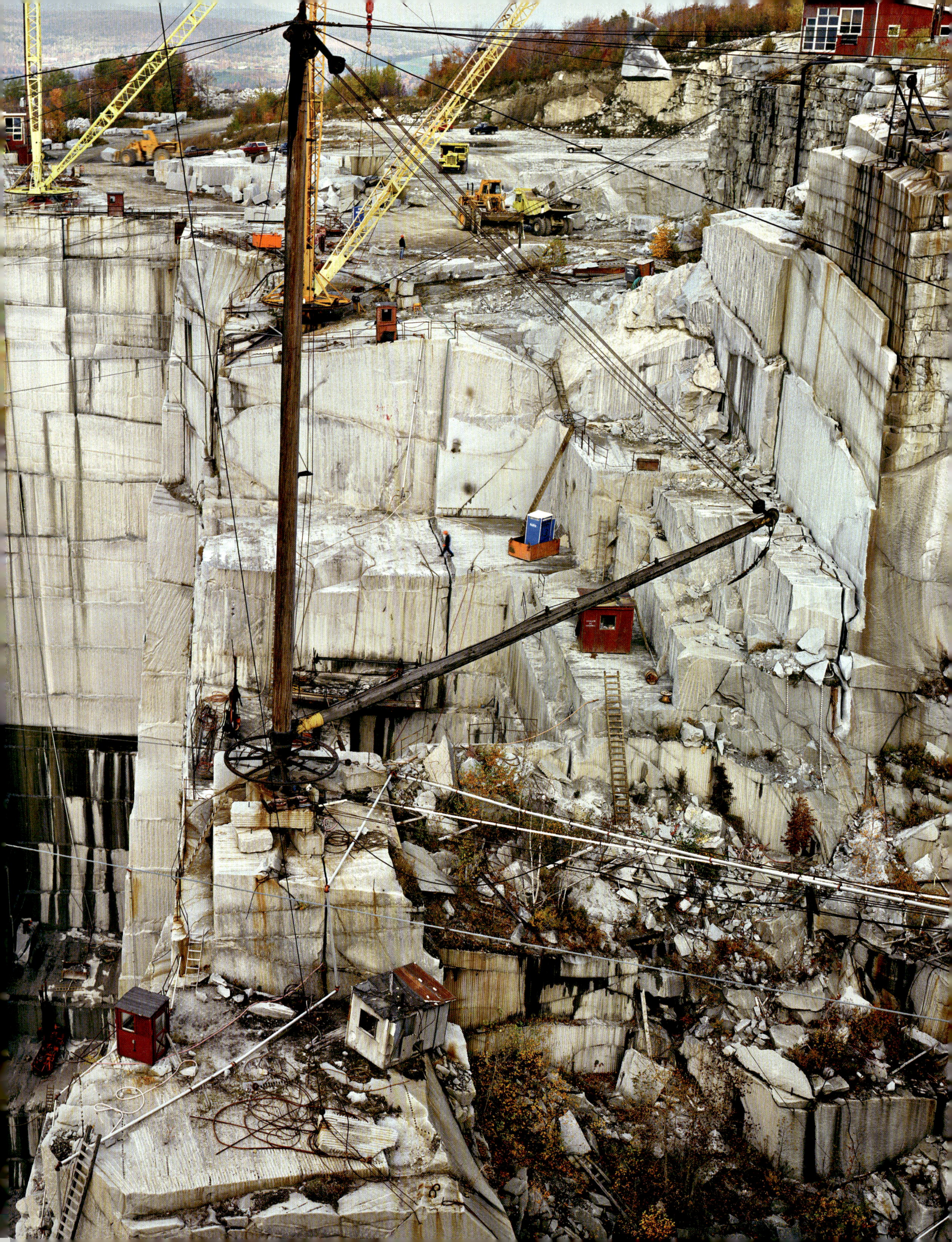

Preface and Acknowledgments

A cemetery provides a unique perspective on the population and social lives of a community, and Hope Cemetery in Barre, Vermont, is certainly no exception: it returns the visitor, in dramatic fashion, to its many powerful moments, but also to a culture and an entire way of life. Amid the more traditional headstones and monuments, including those of people who died during the flu epidemic of 1918–19, are a host of huge and remarkable memorial sculptures—a couple holding hands in bed, a soccer ball, a race car, a replica of Michelangelo's iconic *Pietà*. Then there is the tombstone sculpture of a woman lovingly grasping her dying husband, which provides a clue to the history of this remarkable cemetery and its community. The man depicted here, Louis Brusa, was a stone carver who succumbed at the age of fifty to silicosis, a debilitating and ultimately fatal disease caused by breathing airborne stone dust. These monuments exist, we come to realize, because Barre has been one of the world's preeminent granite quarrying locations for over a century. Brusa's name itself, and those on the majority of the monuments in the cemetery—Orlandi, Chiaravalli, Bianchi, Corti—completes the historical puzzle. At the turn of the last century, Barre attracted immigrants from Carrara, Italy—whose famous quarry has supplied marble for sculptors, including Michelangelo, for thousands of years—to contribute their knowledge and skills to the fledgling industry. Inspiring signs of Barre's still-active quarrying and stoneworking tradition are visible everywhere in the buildings and monuments throughout the town, while the large Italian American population of the area reflects the legacy of the industry.

Over the course of his thirty-year career, Canadian artist Edward Burtynsky, whose work we celebrate in this exhibition and catalogue, has become a pioneering, internationally recognized photographer. Traveling the world, Burtynsky searches out and photographs places that represent the impact that human consumption makes on the earth at its most profound—oil extraction in the United States and in Azerbaijan and shipbreaking in Bangladesh; electronics factories a kilometer in length and immense wire recycling yards in China; nickel and uranium mine tailings in Canada; and marble and granite quarries in, among other locations, Barre and Rutland, Vermont. His photographs of these transformations of nature are breathtaking in size and composition and reflect the consequences that pursuing our material desires has had on the earth. Burtynsky's photographs have been explored in countless publications since 1984, and he has been the recipient of many prestigious awards, including the inaugural TED Prize, *Applied Arts Magazine* book awards, the Officer of the Order of Canada, and, in 2011, the important MOCCA (Museum of Contemporary Canadian Art) award, given for innovation and contributions to the field of contemporary art. He is the founder of Toronto Image Works, a darkroom rental facility, photography laboratory, exhibition space, and digital and new-media education center for photographers of all levels. His work is represented in some of the most important public, private, and corporate collections

Edward Burtynsky
Rock of Ages #19, Granite Section, Rock of Ages Quarry, Vermont, 1991, chromogenic color print. Lent by the Middlebury College Museum of Art, Vermont: Purchase with funds provided by the Walter Cerf Art Fund; 2008.034. (Detail of plate 17.)

in the United States, Canada, and internationally.

Nature Transformed: Edward Burtynsky's Vermont Quarry Photographs in Context takes as its starting point a remarkable group of photographs that the artist took in the Vermont quarries in 1991. A tip from a Vermont quarryman in turn led him, two years later, to some of the oldest and largest quarries in the world—those in Carrara, Italy—and Burtynsky's international pursuit of his conceptual subjects was born. The Vermont quarry photographs, and those from Carrara (as well as India, China, and Portugal), were published in *Quarries,* his first theme-based project and book. *Nature Transformed* reconsiders a selection of these photographs within the context of the history of Vermont's quarry towns—marble in Rutland and its surrounds, and granite in Barre—and the much longer history of the geological formation of northern New England. This exhibition and multi-authored catalogue were co-curated and co-edited by Juliette Bianco, Assistant Director at the Hood Museum of Art, and Pieter Broucke, Director of the Arts and Professor of History of Art and Architecture at Middlebury College and Associate Curator of Ancient Art at the Middlebury College Museum of Art. I wish to thank them for their untiring devotion and superb artistry in creating an exhibition and catalogue worthy of the subject. I wish also to thank the catalogue's essayists: Kirsten Hoving, photographer and Charles A. Dana Professor of History of Art and Architecture at Middlebury College; Ilaria Brancoli Busdraghi, historian and Visiting Professor of Italian at Middlebury College; and Gary D. Johnson, Professor of Earth Sciences, Dartmouth College. Bianco contributed an insightful interview with Edward Burtynsky, and Broucke, the catalogue's engaging introduction.

The interdisciplinary approach of this project is consistent with the purpose of both the Hood Museum of Art and the Middlebury College Museum of Art as teaching museums. For their contributions to the exhibition's presentation at the Hood Museum of Art, I wish to thank Katherine Hart, Associate Director and Barbara C. and Harvey P. Hood 1918 Curator of Academic Programming; Patrick Dunfey, Exhibitions Designer; John Reynolds, Lead Preparator; Matt Zayatz, Preparator; Sue Achenbach, Art Handler; Lesley Wellman, Assistant Director and Curator of Education; Nicole Gilbert, Exhibitions

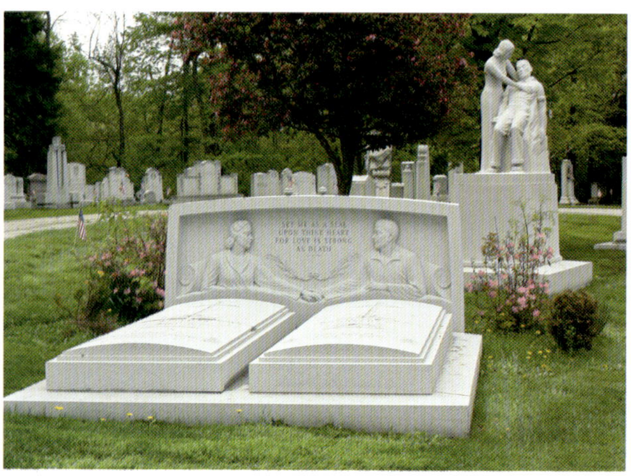

Hope Cemetery, Barre, Vermont.

Coordinator; Kathleen O'Malley, Associate Registrar; Cynthia Gilliland, Assistant Registrar; Nancy McLain, Business Manager; and the entire museum staff. I also thank museum intern Chanon Praepipatmongkol, Dartmouth College Class of 2013, for research related to the catalogue and the exhibition's student programming. I want to specially thank Nils Nadeau, Communications and Publications Manager at the Hood Museum of Art, and Glenn Suokko, independent designer, for the coordination, editing, and design of this elegant volume, which captures the beauty and mystery of Burtynsky's work. For their collaboration in presenting this exhibition at the Middlebury College Museum of Art, I am immensely grateful to Richard Saunders, Director; Emmie Donadio, Chief Curator; Margaret Wallace, Registrar; Ken Pohlman, Museum Designer; Chris Murray and John Houskeeper, Preparators; and Sandra Olivo, Curator of Education.

We would not be presenting this exhibition without Edward Burtynsky's inspiring and enduring photographs, and his interest in and cooperation with this project, which also saw him selecting and printing seven photographs never before exhibited. At Edward Burtynsky's studio, we also want to thank for their collaboration at every stage of this exhibition's preparation Karen Machtinger, Studio Manager and Arts Administrator, and Marcus Schubert, Director of Media, Publications, and Exhibits. At the Aldrich Public Library in Barre, we thank Library Director Karen Lane for researching and making available the historic photographs of life in Barre.

Support of this exhibition came from generous collectors and museum donors, to whom we are most grateful. Raphael and Jane Bernstein, Dennis Sughrue, Daniel Bernstein and Claire Foerster, and the Middlebury College Museum of Art all lent important works to the exhibition. For her generous support of the exhibition's presentation at the Hood Museum of Art, we thank Laurie Jean Weil D.V.M., a valued member of the Hood Museum of Art Board of Overseers, whose contribution was made in honor of her parents, Jean and Bucks Weil, Dartmouth College Class of 1935. We also thank the Marie-Louise and Samuel R. Rosenthal Fund and the Ray Winfield Smith 1918 Fund for supporting the exhibition's presentation at the Hood.

Finally, my most heartfelt thanks go to Raphael and Jane Bernstein, who have enthusiastically supported this project from the very beginning. As supporters of Edward Burtynsky's photographic career for more than twenty years, they most generously funded the design and printing of this handsome catalogue. Their longstanding support of the Hood Museum of Art's teaching mission is appreciated by thousands of Dartmouth students, dozens of faculty members, and the entire museum staff each year.

Michael R. Taylor
Director

NATURE TRANSFORMED

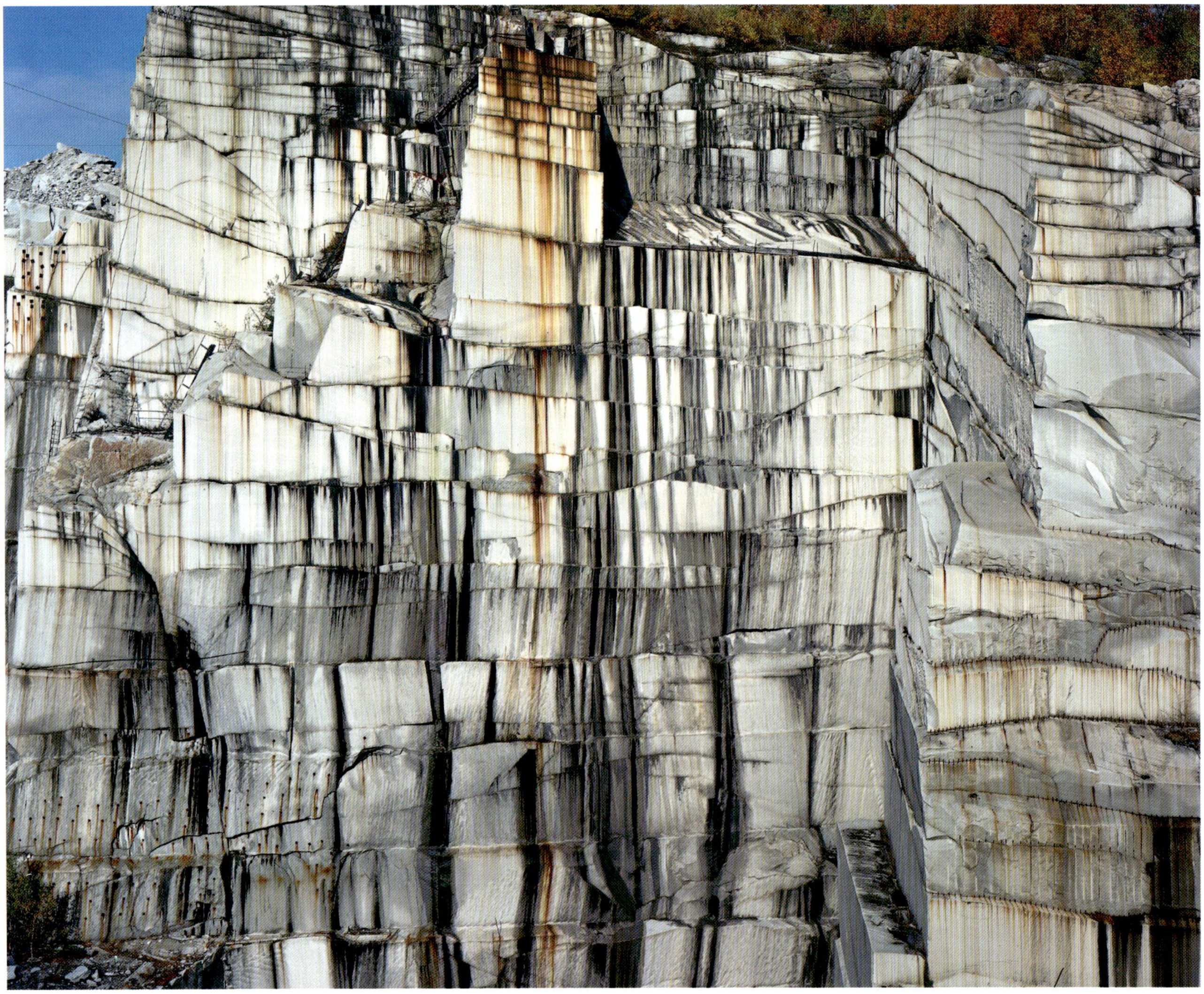

Edward Burtynsky
Rock of Ages #26, Abandoned Section, E. L. Smith Quarry, Barre, Vermont, 1991.

1

Introduction: Platonic Voids—Ecological Scars

Edward Burtynsky and the Quarries of Vermont

Pieter Broucke

I remember looking at buildings made of stone, and thinking, there has to be an interesting landscape somewhere out there because these stones had to have been taken out of the quarry one block at a time. I had never seen a dimensional quarry, but I envisioned an inverted cubed architecture on the side of a hill. I went in search of it, and when I had it on my ground glass, I knew that I had arrived.

—Edward Burtynsky, 2001[1]

In his spectacular large-scale photographs of the marble and granite quarries of Vermont, Edward Burtynsky captures the manmade cavities systematically carved out of the landscape. Steep quarry faces lead the eye from the recognizably "Vermont" landscape on top down into the bowels of the earth, the cavities often terminated by pools of stagnant water. The faces step down in precipitous yet rational patterns, determined by the regular extraction of what Burtynsky (and the quarry industry itself) calls "dimensional stone"—stone quarried for blocks to be used as architectural members or tombstones. Of course, these patterns have been rendered irregular by geological accident, weathering, and mechanical intervention. Seemingly objectively, the resulting dramatic vistas leave the viewer vertiginous, overwhelmed by their sheer power. Burtynsky's photographs capture an essential quality that resonates with the viewer beholding the images of these quarries—that of the truly epic presence.

A series of exploratory field trips to quarries all over his native Canada during the late 1980s frustrated Burtynsky, as none of the quarries he visited met the expectations he had set for his photographs, allowing him to capture on film what he so clearly saw in his mind's eye. It was not until after a quarryman in Ontario informed him he should look at the quarries in Vermont that Burtynsky first came to Barre in the fall of 1991.[2] There, Burtynsky finally found what he had so doggedly sought: truly monumental voids, or "inverted skyscrapers," in his own words, that could serve as the eloquent and visually stunning manifestations of his vision.[3] It is precisely this type of quest that makes Burtynsky a conceptual artist—diligently seeking to realize an idea in tangible form by means of photography—rather than a documentary photographer per se.

The quarries Burtynsky saw first in Barre and later at other venues in Vermont are "inverted skyscrapers" in more ways than one. Whereas skyscrapers normally appear as towering masses surrounded by space, the quarries consist of downward spaces surrounded by mass. Whereas many skyscrapers tend to be Platonic forms, simple and regular geometric solids, several of the quarry spaces could be considered in turn "Platonic voids," crisply delineated spaces with stark geometries (plate 4). Appearing as they do below ground, the quarries are also in a sense "constructed" upside down, beginning at ground level and reaching toward the center of the Earth as time goes on; skyscrapers, of course, rise above the ground, are constructed from the bottom up, and reach for the heavens. The oldest parts of the quarries are at the

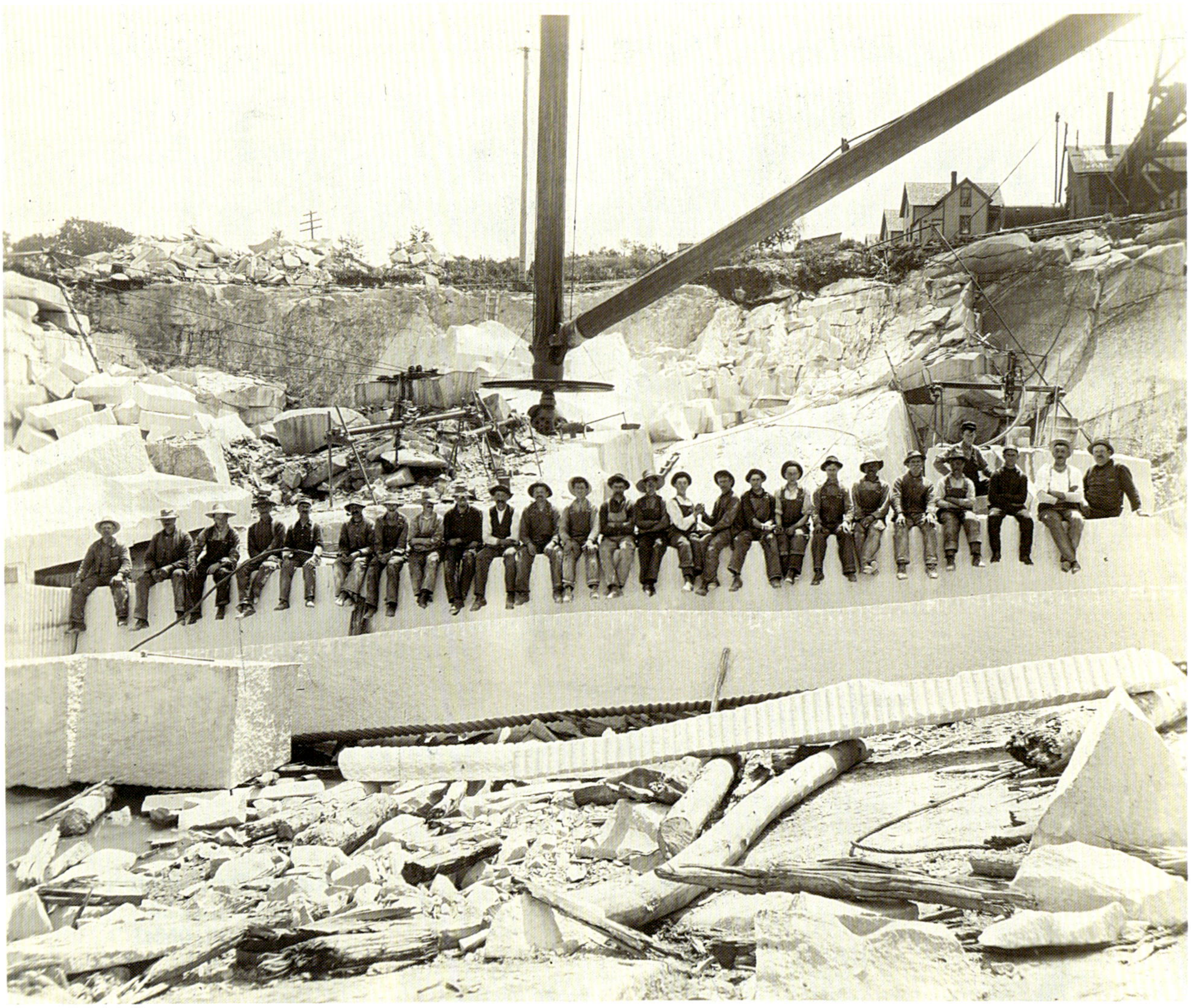

Figure 1.1
Photographer unknown, Wells-Lamson Quarry, about 1908.

top; the most recent parts, at the very bottom. This inverted chronology is also revealed in a general way when one compares images from a century ago, with quarrying occurring much closer to the surface (fig. 1.1; see also fig. 3.1), to Burtynsky's images from just two decades ago, with quarrying taking place deep in the earth (plates 4, 17, and 19). This progression makes for interesting "reading" of the quarry faces themselves as well: the traces left by the processes of extraction changed over time as the processes themselves evolved. Not surprisingly, in many quarries the blocks extracted early on were smaller than those permitted by modern technology, as is clear from the differences in size between the upper facets of the quarry faces and the ones farther down (plate 4). The quarry faces also provide evidence for the changing technologies used in block extraction, from the introduction of steam-powered pneumatic equipment in the late nineteenth century, to parallel drilling to release the blocks from the living bedrock, to the application of wire stone sawing directly in the quarries from the twentieth century onward (plate 17).

The dimensional quarries that Burtynsky photographed are also "inverted" in another, more fundamental sense, namely in their existence as cultural objects. In *The Shape of Time,* published in 1962, archaeologist and art historian George Kubler famously argued that everything manmade might be considered art, because, due to humankind's innate inertia, "nothing gets made unless it is desirable."[4] There is no doubt that the quarries are manmade, of course, and they certainly qualify as architecture (and therefore art), at least as much as the ancient Bronze Age *tholos* tombs in Greece, which were constructed entirely underground and have long been considered architecture as well. But what sets quarries apart from most other architecture, and what renders them a special case in Kubler's theories, is the fact that they themselves were not "desired" as such. What gives the quarries their architectural form is the process of extraction, and what was "desired" from the quarries, to stay with Kubler, is only what was removed from them. As such, quarries are what Burtynsky calls "residual landscapes," manmade but basically leftover cultural products in which appearance—not to mention deliberate aesthetic design—played no role.[5] We owe it to such artists as Burtynsky to make us see these quarries as the literally extra-ordinary (and awe-inspiring) cultural products they are.[6]

Having found the quintessential quarries he had been looking for in Barre, and later all over Vermont, Burtynsky subsequently found more of them elsewhere, in various parts of the world: first in Carrara in Tuscany, Italy; then also in Makrana in Rajasthan, India; in Xiamen in Fujian Province, China; and in several locations on the Iberian peninsula, in both Portugal and Spain. The resulting series of images, taken between 1991 and 2006, was presented in a monumental and momentous publication, *Edward Burtynsky: Quarries,* published in 2007.

From the perspective granted by two decades of hindsight, Burtynsky's quarry project emerges as pivotal to his development as an artist, coming as it did in the wake of a project on North American bank architecture.[7] His deeply conceptual interest in quarries was initially mostly of a formal nature. The very first photograph Burtynsky took in Barre, in the fall of 1991, is telling in that regard (see p. xii). The image prominently features the visual qualities of the nearly vertical quarry face, which resembles a busy patchwork of striped surfaces stitched together. The stone fills the image from edge to edge, with perspectival relief only in the upper two corners: a swatch of blue sky enters at the upper left while a shallow stretch of trees links the monumental quarry face to the distinctive fall foliage of Vermont in the upper right. Only upon closer scrutiny do ladders and stairs become visible, but all they do is remind the viewer of the scale of the quarry. The image concerns itself first and foremost with the formal, spatial, and temporal aspects of the quarry alone.

Following his first visit in 1991, Burtynsky took five additional trips to Vermont to photograph more quarries, both granite and marble, still active and abandoned, in Barre, Rutland, Proctor, Danby, and Rochester. The images indicate the different trips: some feature the above-described foliage of maple trees from the fall of 1991 (plates 13, 15, and 17), others the patches and ribbons of snow on the quarries' ledges in the winter of 1992 (plate 16). Across the whole group of images that resulted from those six Vermont trips, Burtynsky subtly combines his instinctive appreciation for the powerful formal aspects of the quarries with his growing subjective awareness

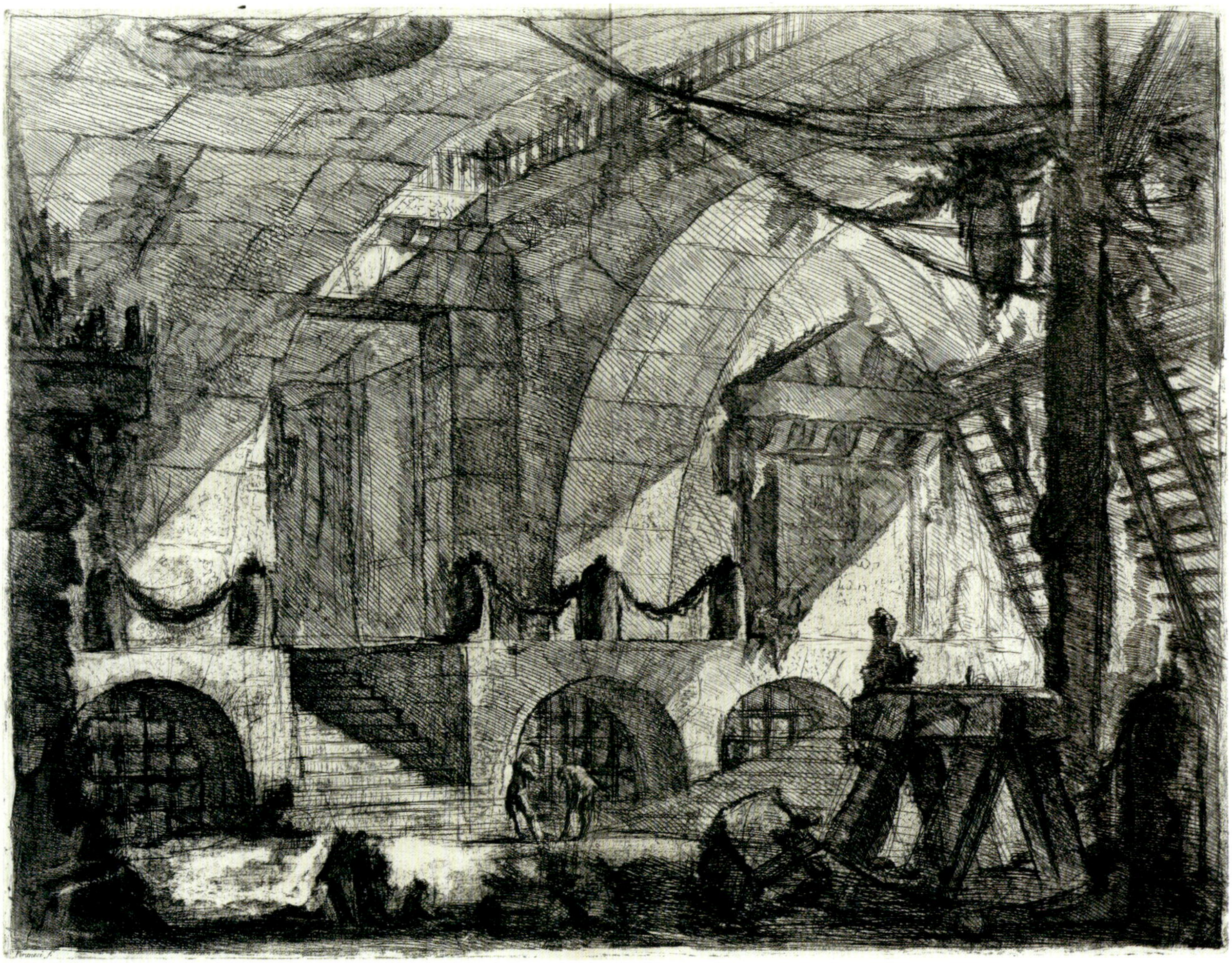

Figure 1.2
Giovanni Battista Piranesi, *The Sawhorse,* from the series *Imaginary Prisons (Carceri d'Invenzione),* about 1745–50, published 1753, etching on heavy laid paper. Hood Museum of Art, Dartmouth College; PR.977.25.57.

of the devastation that large-scale industrial quarrying has wrought on the landscape. The detritus of abandoned equipment, derricks, sheds, stairs, cables, tanks, discarded blocks and stones (referred to as "grout" in the industry), and other litter becomes increasingly prominent in the photographs (plates 13 and 20).[8] The artist's initial impression of the monumental voids as formal presences inserted within the landscape gradually commingles with his recognition of the quarries as ecological wounds inflicted upon the landscape:

> *We are surrounded by all kinds of consumer goods, and yet we are profoundly detached from the sources of those things. Our lifestyles are made possible by industries all around the world, but we take them for granted, as background to our existence. I feel that by showing those places that are normally outside our experience, but very much a part of our everyday lives, I can add to our understanding of who we are and what we are doing.*[9]

Describing what he encountered in the quarries as "normally outside our experience" is no overstatement on Burtynsky's part, and some of the images depict spaces that seem utterly *other*worldly. In particular, the photographs of the underground marble quarries in Danby display, through their depiction of cavernous, seemingly endless space and their dramatic perspective, a haunting and claustrophobic grandeur recalling Piranesi's *Imaginary Prisons* (fig. 1.2, plate 5) or science fiction films (plate 6).[10] Not surprisingly, these enclosed spaces are also off limits to visitors, save for stone industry initiates.

Equally alienating is Burtynsky's photograph of the calcium carbonate waste area associated with a marble processing plant that OMYA, an international mineral processing company, operates in Florence, Vermont, near Proctor (plate 3).[11] OMYA extracts granulate marble from quarries in several Vermont locations, including Middlebury and Danby, and grinds it up into particles finer than dust for subsequent use in paper, paint, and even toothpaste. Processing the marble granulate generates waste, gathered in large dumps, that heavily impacts the environment.[12] Burtynsky frames the compacted waste dump, pocked with troubled puddles, in a manner that evokes a still Arctic landscape, with bergs of ice and powdery snow piled up high in a tranquil frozen sea. While this soft but eerie image of the marble industry contrasts in many ways with the hard-edged residual landscapes of the dimensional quarries featured in the other Vermont quarry photographs, the depicted landscape is equally residual, and equally wounded.

In the series, Burtynsky leaves us with some hope, however—among others, the images of the abandoned marble quarries near Rutland and Proctor, in the Champlain Valley in western Vermont, show how environmental wounds can heal with time, and how nature can overtake and reconquer civilization in the long run (plates 1 and 2). With the passage of almost geological units of time, the quarries begin to resemble geology itself, and the poetic images of these abandoned quarries recall none other than Goethe, the great Romantic author who, upon first encountering the majestic Greek temples in the overgrown ancient city of Paestum, south of Naples, in the spring of 1787, was uncertain whether he was beholding "rocks or ruins"—nature or civilization.[13] Rather than highlighting the epic industrial rationality that is so prominent in his granite quarry images, Burtynsky displays in the images of the abandoned quarries near Rutland a poetic romanticism that is almost consoling regarding nature's fate.

Given the above-described emphases that Burtynsky places on the ecological impact of modern life in his later oeuvre, the relatively early photographs he made of the Vermont quarries, taken together, emerge as a formative body of work that can be considered as pivotal for his artistic career. What started as conceptual and formal in nature ended up as powerful environmental commentary as well.

While many of these photographs have been featured individually as well as collectively in several exhibitions and many museums as the works of art that they first and foremost are, the purpose of *this* exhibition and its accompanying catalogue is to properly contextualize Burtynsky's images of the Vermont quarries, given their important place in his artistic evolution. In order to present a comprehensive and cohesive overview of the artistic material Burtynsky generated for this project, seven previously unpublished photographs are included in the exhibition and published for the first time here (plates 3, 5, 6, 8, 15, 16, and 20). Four Burtynsky photographs of Carrara have been included here as well (plates 7–10), because of the importance of the marble quarries at Carrara for cultural history in general (they have been in continuous operation since the first century B.C.E. and provided marble for the Rome of Augustus and the sculptures of Michelangelo) and for social history in particular (the first Italian stoneworkers who immigrated to Vermont came from that general area). It is no coincidence that the venues of this exhibition, the Hood Museum of Art at Dartmouth College in Hanover, New Hampshire, and the Middlebury College Museum of Art in Middlebury, Vermont, are museums attached to educational institutions and thus well suited to the consideration of Burtynsky's Vermont quarry photographs in a variety of contexts.

In that regard, each of the three essays in the present volume focuses on a discrete aspect of these images: their place within the history of photography; the quarries as lasting monuments to globe-encompassing social history; and the complex geology of Vermont that resulted in its rich array of granite and marble, occurring in a broad range of hues and qualities. All three naturally touch in some way upon the rich history and technology of quarrying itself—the industrial processes responsible for the creation and formal articulation of the quarries as "nature transformed." Given the size of the quarries and the extent of the stone industry in Vermont, many technologies were actually first developed in that state. For instance, when Eben Judd discovered a small pocket of marble of exceptional quality near the waterfalls in Middlebury in 1802, a machine was invented that cut the stone by means of water-powered saws, a technology that was subsequently exported all over North America and into Europe.[14] Later in the nineteenth century, Emery L. Smith, who in 1880 founded the granite quarry in Barre that still carries his name, pioneered the use of the fixed derrick for lifting blocks out of the quarry, detonated explosives by means of electricity for the first time, and first used compressed air to power his drills.[15] These technological developments directly affected the images that Burtynsky captured, as well as the social history and Vermont geology that informs them.

In her essay, Kirsten Hoving situates Burtynsky's Vermont quarry images within his oeuvre, personal experiences, and schooling, while describing how they ultimately led to photographic projects on mining, oil refining, and dam construction the world over. She also assesses Burtynsky's place as a prominent contemporary photographer within the history of photographing quarries as a subgenre of industrial documentary photography. Against this background, she poignantly presents Burtynsky's photographs of the Vermont quarries as documents that are both beautiful and provocative, and that bear witness to how we as a species treat and value our environment.

For her essay, Ilaria Brancoli Busdraghi focuses on the social history surrounding these quarries, as manifestations of the gradual industrialization of quarrying in Vermont that occurred in the period between the Civil War and World War I, a process that was characterized socially by immigration and cultural assimilation. By focusing on the Italians who came to work in the marble and granite industries in Vermont, she presents a case study of how issues of cultural heritage and political identity both shape and are shaped by local conditions of industrial organization. In that regard she draws a sharp line between the social and political aspects of the marble and granite industries in Vermont, which were respectively centered around Proctor and Rutland in the Champlain Valley, on the one hand, and around Barre in the Green Mountains, on the other.

The essay by Gary Johnson explores the rich geology of Vermont, tracing a history of the Vermont rock formations and the primary materials that are being extracted from the quarries: the famous Vermont marble, used in the construction of the Jefferson and Lincoln Memorials in Washington, D.C., and the stone referred to as "Vermont granite." As the author explains in his essay, not all of that stone is a true "granite" in scientific terms. Even "Barre granite," a registered trademark, is strictly speaking a granodiorite, though it nonetheless possesses all the desired qualities of granite as such.

In her interview with Edward Burtynsky, Juliette Bianco invites him to consider the Vermont quarry photographs as their author, and he positions the quarry project within his career as an artist forever in search of images that affect our collective consciousness and that are archetypical in a Jungian sense. The connections forged by the Italians who came from Carrara to Barre prompted the artist to go from Vermont to Italy, resulting in the first photographs he took beyond the shores of North America. That Italian project, the direct outcome of his work in Vermont, paved the way for an artistic oeuvre that now encompasses the entire world and led directly to his current work on industrialization in China and on water and agriculture in, among other places, Spain. Burtynsky also muses on the function of photography beyond the artistic—as a medium that allows us to stop time, so to speak, in order to take stock and reflect, and that therefore raises consciousness in a variety of realms, from the formal and architectural to the environmental and political.

Whether Burtynsky shows us the monumental spaces as epic voids (the result of heroic effort) or as scars on the landscape of old and new environmental wounds (the result of ecological mismanagement), his images of the quarries of Vermont confront us with two sides of ourselves, and it may well be that quality that makes them, in the end, so powerful and important. As Burtynsky notes, "There are many things that the rock face can tell us about ourselves."[16]

Notes

1. Edward Burtynsky, "Artist Statement," in *Edward Burtynsky Quarries: The Quarry Photographs of Edward Burtynsky* (Göttingen: Steidl, 2007) [hereafter "*Quarries*"], 9.
2. Personal communication with Edward Burtynsky, Toronto, February 10, 2011.
3. Edward Burtynsky, "Artist Statement," in *Quarries*, 9.
4. George Kubler, *The Shape of Time: Remarks on the History of Things* (New Haven: Yale University Press, 1962), 1.
5. "Residual landscapes" is a term that Burtynsky uses throughout his career to denote the content of much of his work, from the quarries to his more recent projects.
6. Burtynsky himself refers to the notion of desire in the "Artist Statement," where he writes: "The rock face is imprinted with that order, with our methodologies, our desire, our need" (*Quarries*, 9).
7. Personal communication with Edward Burtynsky, Toronto, February 10, 2011. See also "Acknowledgments," in *Quarries*, 190.
8. "Grout" is originally a Scottish word for waste, used in the industry to denote "waste stone." See Todd Paton, *The Rock of Ages Story* (Barre, Vermont: Rock of Ages Corporation, 2003), 28.
9. "Artist Statement," *Quarries,* 9.
10. Giovanni Battista Piranesi, *Carceri d'invenzione (Imaginary Prisons),* a print series published in 1750 and 1761.
11. OMYA acquired the processing plant in 1976 from the Vermont Marble Company, which is discussed in Ilaria Brancoli Busdraghi's essay in this catalogue. See the "OMYA in Vermont" website (http://www.omyainvermont.org; accessed November 12, 2011).
12. In 2002 a group of Vermont community members created "Residents Concerned about OMYA," an organization that has pursued environmental impact litigation against the company. Information on these developments comes from the website of "Vermonters for a Clean Environment" (http://www.vce.org; accessed November 12, 2011).
13. Johann Wolfgang von Goethe, *Italian Journey, 1786–1788,* transl. W. H. Auden and Elizabeth Mayer (San Francisco: North Point Press, 1982), 209, entry for March 23, 1787.
14. Nancy Rucker, "Industry in Early Middlebury," and Polly Darnell, "Henry's Collection," *Sheldon Museum Newsletter* (Middlebury, Sheldon Museum) 5, no. 1 (Summer 1984): 2 and 3, respectively. The dimensional quarry in Middlebury is long exhausted, but the "Marble Works" near the falls—sheds constructed of leftover marble blocks—testify to the town's industrial past. As discussed above, a granulate marble quarry is still in use in Middlebury, to the southeast of the downtown area.
15. Paton, *The Rock of Ages Story,* 27.
16. "Artist Statement," *Quarries,* 9.

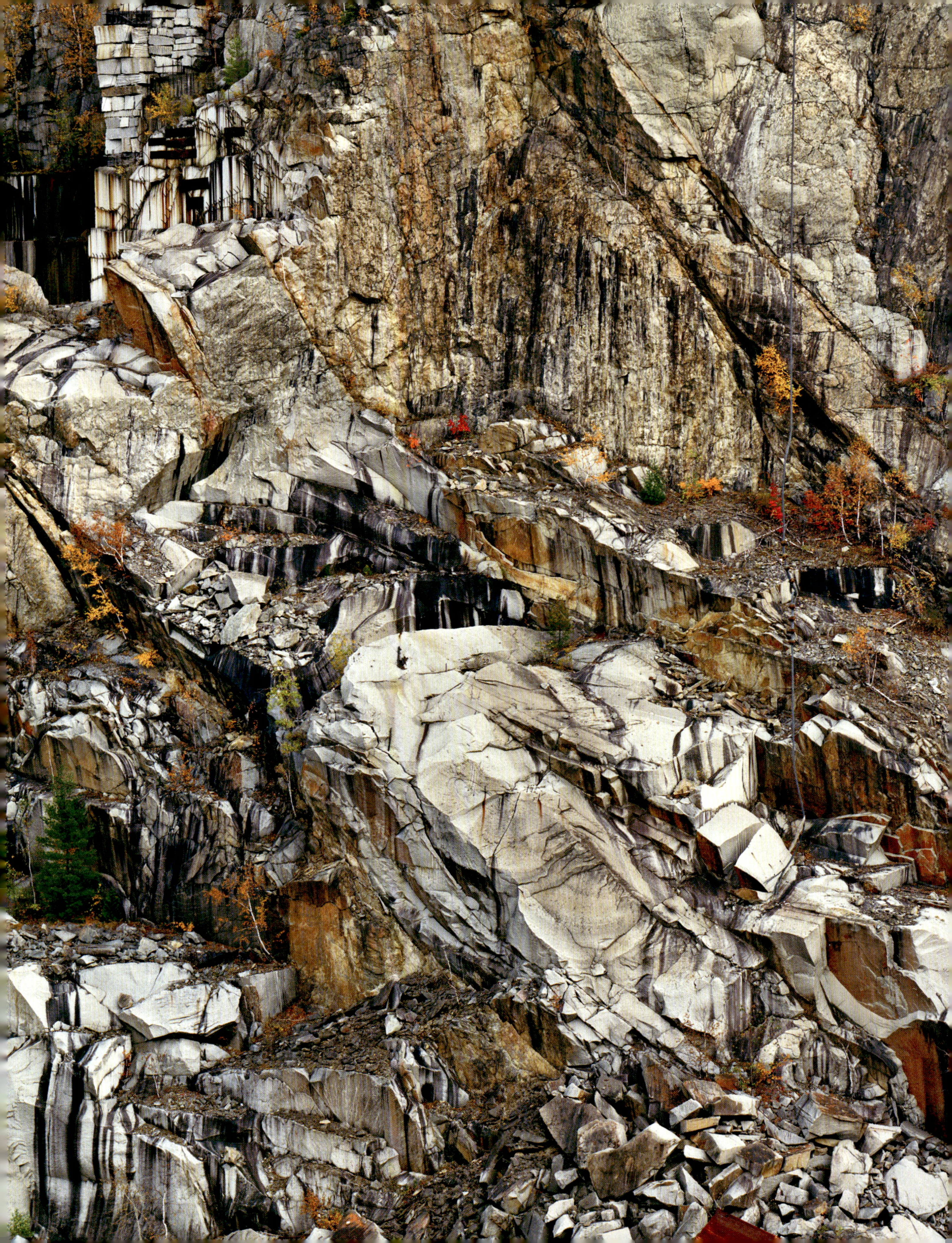

2

Digging Deep

Edward Burtynsky's Vermont Quarry Photographs

Kirsten Hoving

In 1991, Edward Burtynsky made the first of six trips to Vermont to photograph granite and marble quarries. The resulting series launched the career of one of the foremost Canadian photographers of his generation. Combining an environmental ethos with the aesthetics of modernism, Burtynsky produced a body of work in Vermont that chronicles the industrial landscape in images that are lusciously beautiful and ethically disturbing. Featuring vertiginous quarry walls and teal green pools of water, as well as the workers, tools, and detritus of the stone extraction industry, Burtynsky's Vermont quarry photographs combine traditional ways of depicting the industrial landscape with modern concerns about the effects of consumerism upon resource depletion. Eventually expanding this project to include sites in Italy, India, China, Spain, and Portugal, Burtynsky would apply the lessons he learned in the quarries of Vermont to a global context.

Born in 1955 in St. Catharine's, Ontario, Burtynsky studied photography at Ryerson University in Toronto, where an assignment to photograph the human presence in the landscape resonated with his personal experience of growing up in the shadow of a General Motors forging plant, where his father worked until his death at forty-six. Experiencing Toronto's urban landscape led Burtynsky to question the origin of the materials used to build the huge skyscrapers around him: "I felt that Newtonian law implied a reciprocal action in nature—a hole in the ground that meets the scale of the rising of the skyscrapers—and my task was to go in search of the evidence of that reciprocal action, to see what the residual world looked like."[1] A chance encounter in 1981 with an otherworldly landscape of coal mining slag heaps in Frackville, Pennsylvania, opened Burtynsky's eyes to the expressive power of the large-scale industrial landscape. Furthermore, careful observation of nineteenth-century topographical photographs by such masters as Carleton Watkins taught him that their elevated viewpoint could turn a vast expanse of land into what Burtynsky has called "a mythic space, an archetypal sense of the landscape."[2]

An important aesthetic epiphany came in 1982

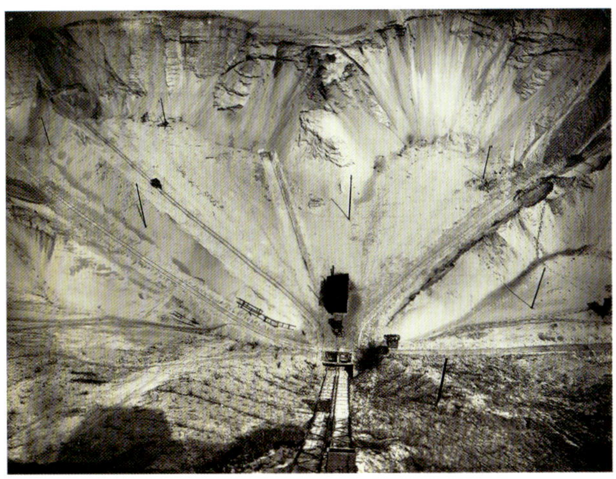

Figure 2.1
August Sander, *Quarry Pit*, about 1925–35, gelatin silver print, $6\frac{9}{16} \times 9\frac{15}{16}$ in.
The J. Paul Getty Museum, Los Angeles

Edward Burtynsky
Rock of Ages #6, Abandoned Section, Rock of Ages Quarry, Barre, Vermont, 1991, digital chromogenic color print. Courtesy of the artist. (Detail of plate 13.)

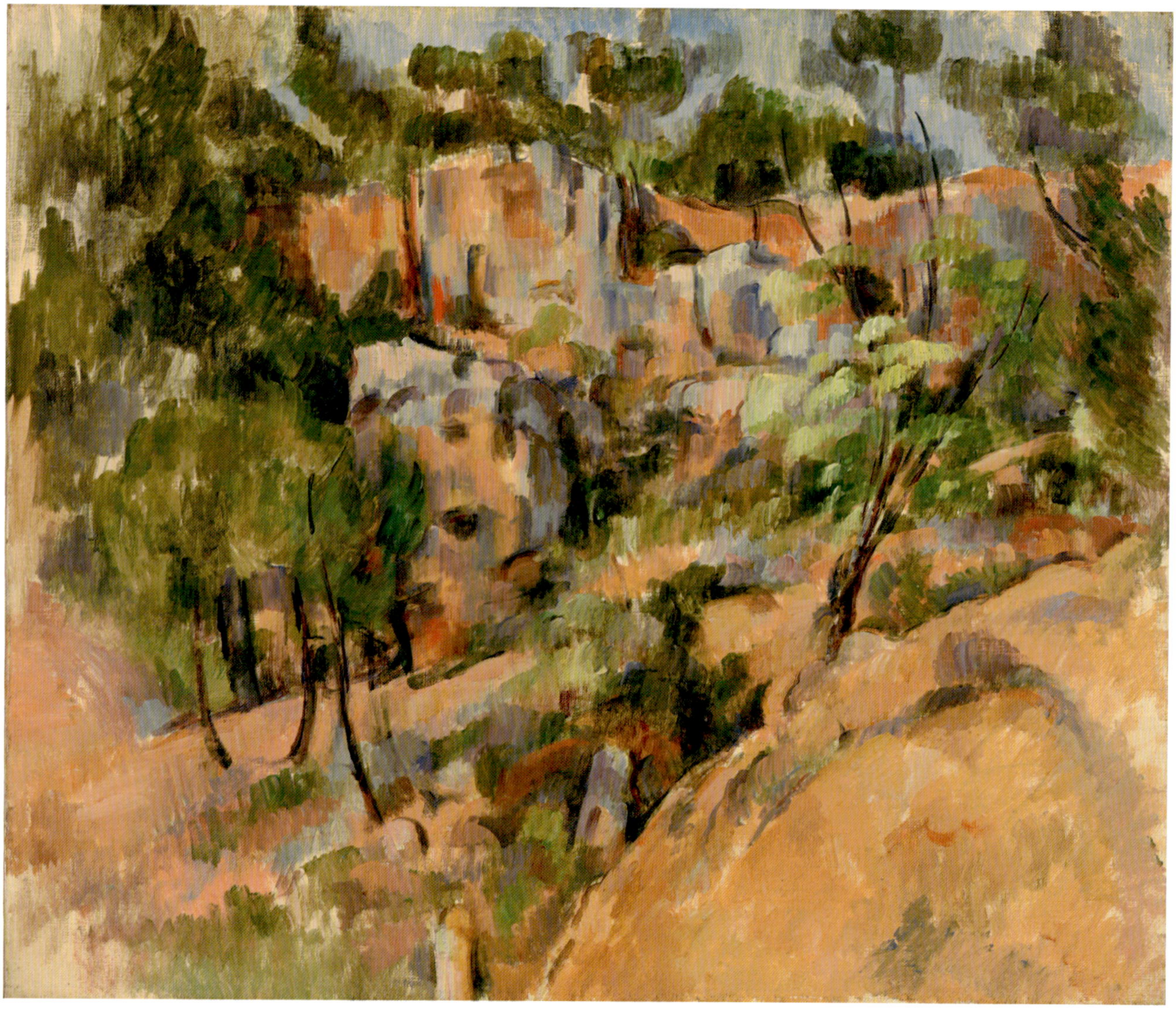

Figure 2.2
Paul Cézanne, *Bibémus*, 1894–95, oil on canvas, 18¼ x 21⅝ in. Photograph © 2011 The Barnes Foundation

at the National Gallery of Canada, when Burtynsky saw an unusual photograph by the German portraitist August Sander (fig. 2.1). Taken near Cologne, the photograph offers a view from above into the depths of a quartz quarry. Using the popular modernist device of creating an abstract pattern by looking directly down on a scene, Sander produced an astonishing view of an industrial landscape that seems turned on its side. Burtynsky recalled: "I stared at it for a minute before I could begin to figure out what I was looking at, and a minute is like eternity to just comprehend what you're seeing. No one had ever done that to me. It was exciting."[3] Sander's photograph showed Burtynsky how an unusual perspective could make a scarred landscape both visually disturbing and compellingly beautiful.

Burtynsky was not alone in his desire to depict signs of the human impact on the natural landscape. At the end of the nineteenth century, Post-Impressionist artists were drawn to quarries as a subject that could express their growing discontent with industrialization. Vincent Van Gogh, Paul Gauguin, and especially Paul Cézanne produced quarry paintings that depicted the intersections between the natural world and the effects of human intervention upon it.[4] In particular, in his many paintings of the abandoned Bibémus Quarry, Cézanne explored the cubic shapes and ochre color of the site, flattening the space into a tightly conceived, purely graphic design (fig. 2.2). More significant for Burtynsky himself was a landmark exhibition curated by William Jenkins in 1975 at the George Eastman House in Rochester, New York, called *New Topographics: Photographs of a Man-Altered Landscape*. The exhibition featured work by an older generation of photographers, including Robert Adams, Lewis Baltz, Frank Gohlke, Stephen Shore, and Bernd and Hilla Becher, whose approach would pave the way for Burtynsky. Eschewing the veneration of pristine wilderness popularized by Ansel Adams and Eliot Porter, New Topographics photographers like Robert Adams produced coolly ironic views of industrial or suburban encroachments on the natural world, such as his *Santa Ana Wash, Next to Norton Air Force Base, San Bernadino, California*, a desert landscape with a faint airplane in the upper-right corner (fig. 2.3). For the most part working in black-and-white with square, medium-format or large-format view cameras, these photographers opened a debate that still rages today about the signs of human presence in landscape photography. Their detached, almost scientific style of depicting the "man-altered landscape" had a significant impact upon Burtynsky and many other photographers of his generation.[5]

In his search for an "inverted architectural" counterpart to the towering stone-clad skyscrapers found in twentieth-century cities, Burtynsky began photographing quarries in Quebec; however, these sites were too small to satisfy his vision of a hole dug deep enough to match the tall buildings of modern Toronto. In 1991 he followed the advice of quarrymen to visit Vermont. There he found stunningly deep granite and intimate marble quarries that fit his needs. He photographed active and abandoned granite quarry sites owned by the Rock of Ages Corporation in Barre, Vermont, abandoned marble quarries near Rutland, an active former Vermont Marble Company quarry in Rochester, Vermont, and the massive underground marble quarries in Danby, Vermont, now operated by the Vermont Quarries Corporation.[6] Reflecting the variety of sites chosen, the mood of these photographs ranges from wonder at the sublime scale of modern extraction activities to elegiac meditations on the eventual return of these sites to something approximating their natural state.

Burtynsky's quarry photographs share a variety of stylistic characteristics. Shot with a large-format camera on a tripod, the eight-by-ten-inch negatives yield precise detail that can be maintained even when the prints measure as much as five feet on the longest side. Views extend from one side of the quarry to the other or to the quarry floor, necessitating the use of a long-focal-length lens that compresses depth, along with a small aperture to retain depth of field. Initially composed upside-down on the camera's ground-glass back, the resulting photographs are masterful graphic designs that can be appreciated as nearly abstract studies in line, light, and texture. At the same time,

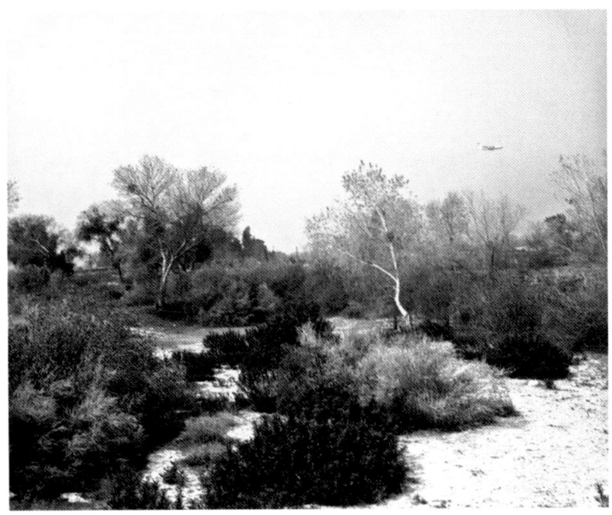

Figure 2.3
Robert Adams, *Santa Ana Wash, Next to Norton Air Force Base, San Bernadino, California*, 1978/82, gelatin silver print, 9 x 11 in. Collection of Middlebury College Museum of Art, Vermont: Purchase with funds provided by the Christian A. Johnson Memorial Fund; 2001.005. © Robert Adams, courtesy Fraenkel Gallery, San Francisco, and Matthew Marks Gallery, New York

however, their sharp detail provides a clear record of the technology of quarrying and its aftermath.

Aesthetic appeal overshadows technological information in Burtynsky's photographs of active quarries, such as *Rock of Ages #7, #19,* and *#39* (plates 14, 17, and 19), all shot at the Rock of Ages Corporation's E. L. Smith Quarry in Barre, Vermont, which, at over five hundred feet deep, is reportedly the largest deep-hole dimension quarry in the world.[7] Seen from afar, the workers are dwarfed by the rock faces that tower above them, and the pneumatic drills, diamond-wire saws, cutting torches, diesel generators, steel cables, ladders, guy wires, portable toilets, handrails, temporary platforms, sheds, rider boxes, derricks, hoist houses, and hoses are reduced to miniature details in the vast panoramas of blocks of rock. In all of his scenes of active quarries, Burtynsky by necessity photographed them from above, producing complex and even confusing views of the industrial landscape. Like August Sander before him, Burtynsky capitalized on the abstraction inherent in this point of view by almost always eliminating the horizon and sky that could help place the scene in a larger, more easily grasped perspective.

Burtynsky is not the only photographer to have taken this approach to photographing quarries. A postcard printed about eighty years earlier shows many of the same devices at work (fig. 2.4). Tiny figures stand on huge blocks while ladders, steam drills, chains, and tools help establish a sense of scale. However, the card's caption, proclaiming this to be "the most scientifically worked quarry in America," sounds a note of technological bravado that is absent from Burtynsky's photographs, where he specifically avoids taking sides, neither extolling the virtues of modern industry nor condemning it outright. Rather, he organizes the quarry's chaos, noise, dirt, and debris into an aesthetic experience of balanced color and forms against the overwhelming backdrop of the huge scale of modern quarrying. Indeed, that physical scale acts as a subtle metaphor for the scale of our unbridled consumption of resources in the modern world—an ongoing concern in Burtynsky's current work that has its origins in the Vermont quarry photographs.

Burtynsky's subjects and style are driven by personal artistic drives and shaped by the loss of his father to the toxic industrial environment in which he worked, but they are also the result of the social and environmental concerns of the late twentieth century. Three-quarters of a century earlier, the scale of industry meant something quite different. A postcard published in 1912 presents the largest boulder ever uncovered of dark Barre granite, "containing 500,000 Cubic Feet, which is 90,000,000 Ninety Million lbs or 45,000 tons." The photograph shows three hundred quarrymen sitting on the boulder, whose blocks eventually filled 1,728 freight cars (fig. 2.5). In this case, scale is measured in numbers, with humans used in the service of boasting the vast scale of consumption

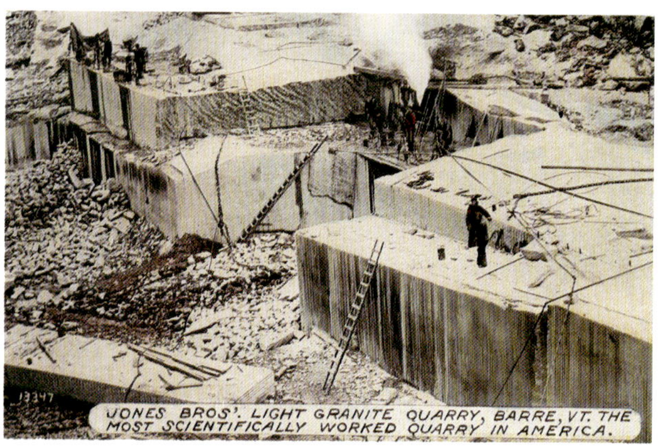

Figure 2.4
Jones Bros' Light Granite Quarry, Barre, VT,
published by the Acmegraph Co., Chicago, about 1908–18. Author's collection.

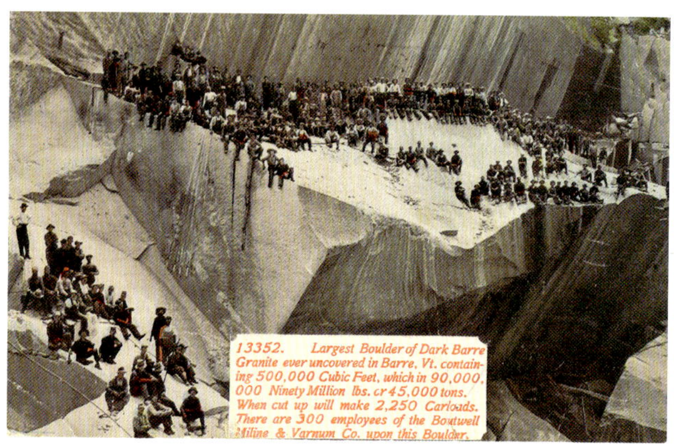

Figure 2.5
Largest Boulder of Dark Barre Granite,
published by the Acmegraph Co., Chicago, 1916. Author's collection.

made possible through their efforts.

Rather than proclaiming the magnitude of industrial achievement, Burtynsky's photographs underscore the magnitude of human consumption: around the time Burtynsky was photographing in Vermont, more than 700,000 cubic feet of granite blocks were produced by the Barre quarries annually. Of those blocks, 85 percent was waste material ("grout"), which was discarded in huge piles in the woods around the quarries, not included in Burtynsky's photographs.[8] Although the photographer has noted that he often looks for the largest example of something—the largest mine, the largest quarry, the most active area—his ultimate goal is to "embody a poetic narrative of the transfigured landscape and the industrial supply line and what that means in our life."[9] Thus, instead of inducing awe through literal numbers—a kind of statistical sublime—Burtynsky's photographs make their point through scale relationships between the humans with their industrial tools and the deep holes dug to extract stone. The artist has observed:

> *Often my approach, the compression of space through light and optics, also yields an ambiguity of scale. You don't know how large anything is until you see a detail whose size you can recognize. Then you have to reconstruct reality to comprehend how big the place really is . . . We need to put our human perspective into these images, and our presence is dwarfed by the spaces we've created. It's an interesting metaphor for how technology seems larger than life, larger than our own lives.*[10]

Fifty years before Burtynsky photographed in Vermont, Farm Security Administration (FSA) photographer Jack Delano visited some of the same quarries in September 1941.[11] As one of the many New Deal programs formed in the wake of the Great Depression, the FSA's Historical Section documented the effects of poverty and the eventual economic recovery by, in the words of director Roy Stryker, "introducing America to Americans." In the early 1940s, technology did not seem "larger than life"; instead, it offered a way out of economic decline. Accordingly, Delano photographed workers transported to the quarry in a railroad car, close-ups of derricks (fig. 2.6), and views of workers and buildings at the tops of the quarries. What Delano communicated in his photographs of these sites was a story about land being worked on a mammoth scale in the name of production by heroic men whose mastery over the rock and the machines used to move it signaled the country's return to economic prosperity. But even Delano was seduced by the possibilities for graphic design and modernist abstraction in at least one of his photos, a strange view looking down into the Wells-Lamson Quarry with shapes formed by the contrast between dark shadows and brightly lit areas (fig. 2.7).

In addition to photographing active quarries, in 1991 Burtynsky also documented quarries that were no longer being worked. In some cases, as in *Rock of Ages #4* and *Rock of Ages #45* (plates 12 and 21), he was

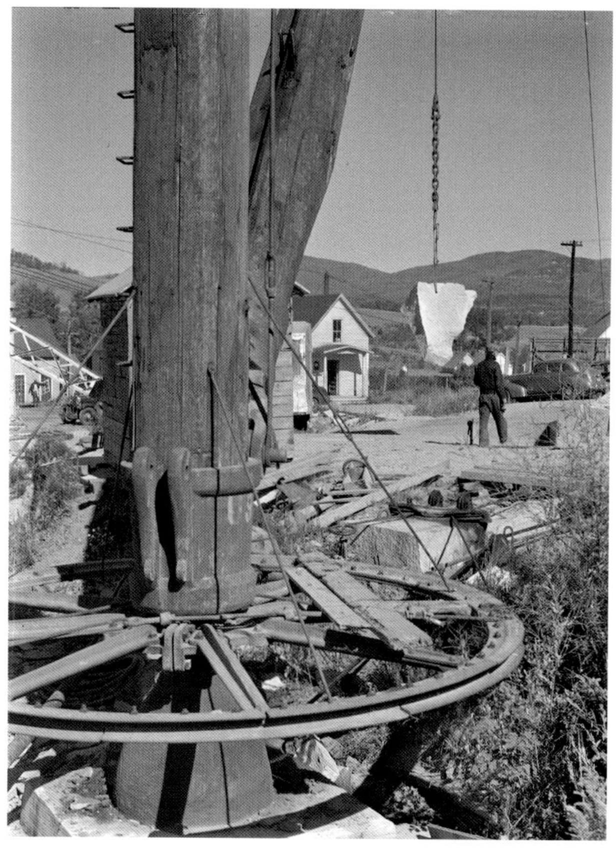

Figure 2.6
Jack Delano, *At the Whitmore* [sic] *and Morse granite quarry in East Barre, Vermont*, September 1941. Library of Congress, Prints & Photographs Division, FSA-OWI Collection, LC-USF34-045980-E.

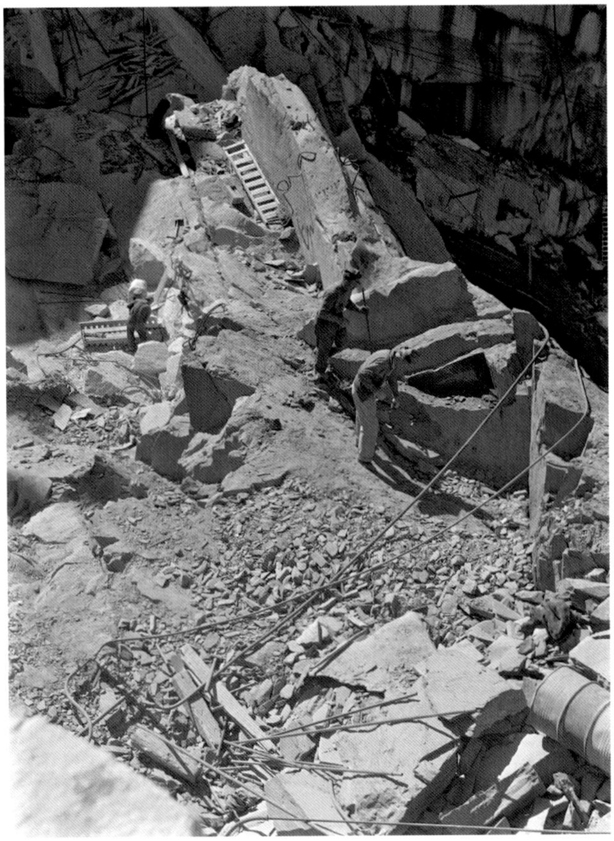

Figure 2.7
Jack Delano, *At the Wells-Lemson [sic] granite quarry near Barre, Vermont*, September 1941. Library of Congress, Prints & Photographs Division, FSA-OWI Collection, LC-USF34-045961-D.

drawn to the linear pattern of the blocks and the dark vertical stains left by the water used for dust control and cooling in wet-drilling. He also included other manmade elements that remain as a kind of industrial graffiti charting what happened to the rock: the heads of huge bolts that were used to stabilize the rock face, as well as the wedges and ladders left behind that constitute what Burtynsky has since called the "patinas" of the walls.[12]

Many of Burtynsky's photographs of inactive quarries, such as *Rock of Ages #12* and *Rock of Ages #25* (plates 15 and 18), incorporate the rectangular pools of water found at the bottoms of most quarries. Strange reminders that long ago all of Vermont lay beneath a shallow sea, the pools are formed by accumulated underground water, water pumped from other quarries, and rain; they are turned deep teal blue or bright turquoise by the algae that feed on the mineral deposits in the water.[13] Although deceptively tranquil, the quarry pools are often hundreds of feet deep, with ledges just beneath the surface still holding derricks, ladders, guy wires, sheds, and other industrial detritus too cumbersome to pull out. In some cases, such as *Rock of Ages #6* (plate 13), the scene Burtynsky showed us in 1991 is largely under water today.

These are the "residual" landscapes most typical of Burtynsky's ongoing engagement with the industrial scene, transfigured sites that are elegiac remnants of human commercial interaction. In some of the inactive quarry scenes there is a mood of apocalyptic destruction, with abandoned sheds and cables strewn among the huge boulders that litter the rock face. In others, discarded ladders suggest escape from some earlier cataclysm or the remnants of long-gone cliff dwellers. Trees and plants signal natural processes of reclamation that will eventually hide nearly all evidence of human intervention under water. However, as Burtynsky's photographs of deep-hole quarries remind us, the material that was removed will not be replaced. While a Rock of Ages Corporation tour guide recently assured visitors that 4,500 years' worth of granite quarrying remains in Barre, there is no denying that the rock that has already been extracted is gone for good.[14]

This note of loss also infuses Burtynsky's photographs of marble quarries located in Rochester, Danby, and near Rutland, Vermont. *Vermont Marble Company #52* (plate 4) pictures the Vermont Verde Antique marble quarry in Rochester, a two-hundred-foot-deep quarry mined from the floor. Seen from the observation deck above, the deserted quarry is traversed with stairways and platforms that lead to the bottom, where yellow hoses and belt saws form abstract designs. The somber mood of this reverse-ziggurat-shaped hole, reinforced by the dark green stone and lack of human presence, answers Burtynsky's imperative of finding reciprocal negative spaces on the land that match the skyscrapers filling modern cities.

Two further photographs, *Marble Quarry #001* and *Abandoned Marble Quarry #18* (plates 1 and 2), are slightly different views of the same abandoned marble quarry near the city of Rutland. The first is an uncharacteristic close-up of the beautiful pattern of colors and textures of a worked rock face, while the second shows that same rock face at an oblique angle, with the view widened to reveal the stepped shelves left over from quarrying as they drop toward a dark pool of water. How different this image is from a nearly identical rock formation photographed by Carlos W. Nichols for a nineteenth-century stereograph (fig. 2.8). Whereas Nichols documented the ladders, ropes, and quarry blocks we are accustomed to finding in Burtynsky's photographs as well, Burtynsky documents a transformed post-industrial landscape where trees and floating leaves suggest a reversion to an earlier, preindustrial condition. This reversion is, of course, an illusion, since hidden beneath the water lies the evidence of the site's former use.[15] Still, there is a hint of redemption in a naturally recycled landscape where a quiet forest pool ringed with autumn trees so poetically masks what remains beneath.

That note of redemption is rare in Burtynsky's work. Central to his artistic expression is a reminder that there are consequences to the scale of modern consumption, consequences that have irreversible effects on our environment and our future. He observes: "Underneath every picture is the fact that nature is being pushed back and our footprint is just getting bigger."[16] Elsewhere, he states: "Ultimately, what I'm looking for are interesting places and moments to embody my poetic narrative of the transfigured landscape, the industrial supply line and what that means in our life."[17] Despite their aesthetic appeal, Burtynsky's photographs urge us to ask important questions about where the materials come from for the manufactured objects that we use and discard.

Burtynsky's quarry photographs belong to a tradition that Perry Miller, David Nye, and others have dubbed the "technological sublime."[18] The sense of wonder, awe, and even terror embedded in large-scale technological efforts like hydraulic mining, a hydroelectric dam, or the atom bomb can be seen in photographs by artists as varied as Carleton Watkins (fig. 2.9), Margaret Bourke-White, and Harold Edgerton. While each of these artists conveys a feeling of awe at the overwhelming size and scale of human technological achievement, Burtynsky tempers that awe with a restrained and literally distanced viewpoint. Indeed,

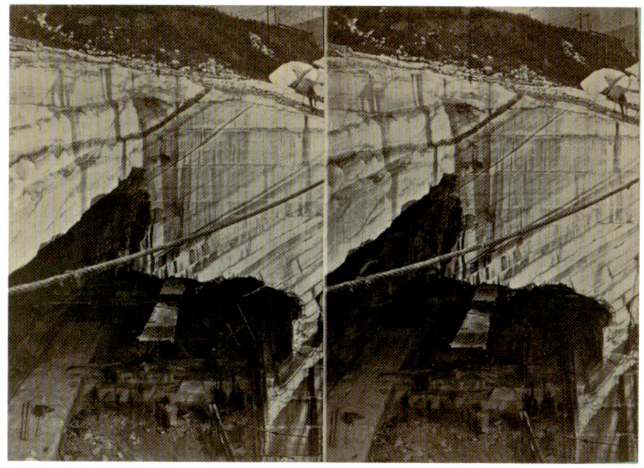

Figure 2.8
Carlos W. Nichols, *Vermont Marble Company's Quarries, West Rutland, VT*, about 1865–85. Mirian and Ira D. Wallach Division of Art, Print and Photographs, New York Public Library.

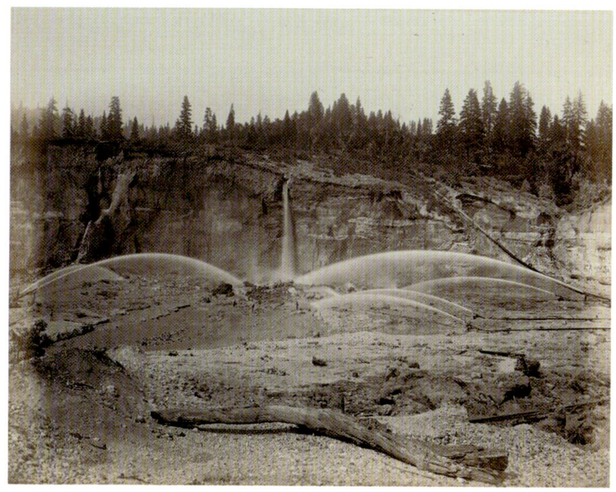

Figure 2.9
Carleton E. Watkins, *Malakoff Diggins, North Bloomfield, Nevada County, CA*, about 1871, albumen print, 15 3/4 x 20 5/8 in. San Francisco Museum of Art: Purchase through a gift of the Judy Kay Memorial Fund.

his photographs may strike viewers as ambivalent: on one hand, the quarry images are graphically enticing in their color and design; on the other hand, they are chilling examples of human undoing of the natural order. When he takes us underground to the Vermont

Quarries Corporation Danby Imperial marble quarry (plates 5 and 6), however, we enter a less ambiguous place that is brooding, claustrophobic, and frightening. Unlike his other quarry photographs, with their cool, restrained elegance, the intense-lit underground areas surrounded by deep shadows invoke nothing so much as William Blake's "dark Satanic Mills."

In the past, photographers found the human toll exacted by the quarrying industry to be especially horrific. For example, in 1910, Lewis Hine photographed illiterate children at work in the marble sheds of the Vermont Marble Company for the National Child Labor Committee (fig. 2.10). Eventually, child labor laws kept children out of factories and more rigorous workplace laws protected stonecutters from the scourge of silicosis. In more recent years, concerned photographers like the Brazilian photojournalist Sebastião Salgado have continued Hine's approach by concentrating on the human dimension of labor, particularly in developing nations. However, for Burtynsky, today's problems extend beyond a particular individual, age group, or national boundary. The workers, presented in the quarries as tiny figures in vast settings when they appear at all, are faceless cogs in a huge industrial machine designed to satisfy our voracious appetites for manufactured goods. There is a human dimension to Burtynsky's work, but it includes all of us, since as consumers we are implicated in a cycle of production, use, and waste disposal of goods and resources that cannot continue indefinitely.

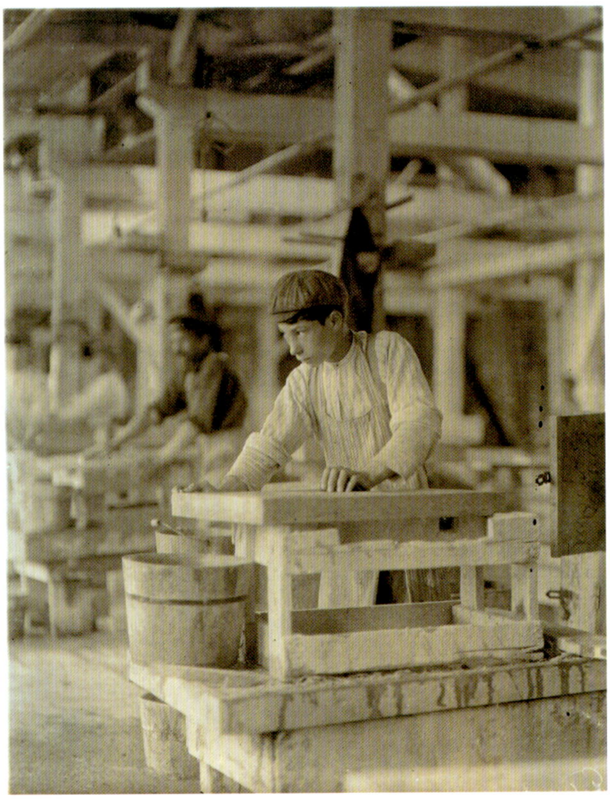

Figure 2.10
Lewis Wickes Hine, *Young boy polishing marble in Vermont Marble Co., Centre Rutland, Vt. Most of these are illiterates. Location: Centre Rutland, Vermont*, September 1910. Library of Congress, Prints & Photographs Division, LC-DIG-nclc-04580.

Burtynsky claims his photographs are not meant as political or environmental diatribes, observing:

> *We are compelled to progress. We have extracted from the land from the moment we stood on two feet. The entire twentieth century has been a revving up of this large consumptive engine. It's not a question of whether we are going to stop consuming. It's not going to happen . . . The only thing we can do tomorrow that is different from today is to manage what we are doing in a better way. We happen to be very successful, and that may eventually be our undoing, because we are also very destructive. We must learn to be more conscious custodians of the resources that we have been given.*[19]

With photographs of Vermont quarries that are both beautiful and provocative, Burtynsky challenges viewers to look beneath the surface at the larger issues facing us on a global scale. Since working in Vermont, Burtynsky has photographed mines in Montana, nickel tailings in Ontario, piles of tires in California, oil refineries in New Brunswick, shipbreaking in Bangladesh, and factories, cities, and dam construction in China. With each project his mission has become clearer: to help people understand that we are "6.5 billion humans living off a precariously balanced, finite planet."[20] Each of Burtynsky's photographs makes an important statement about our values, our excesses, and our environment, and, while delighting the eye, they instruct the mind about the choices that face us today and tomorrow.[21]

Notes

1. Michael Torosian, "The Essential Element: An Interview with Edward Burtynsky," in *Manufactured Landscapes: The Photographs of Edward Burtynsky*, ed. Lori Pauli (Ottawa: National Gallery of Canada, in association with Yale University Press, 2003), 49.
2. Ibid., 46, 55.
3. Ibid., 52.
4. See, for example, Paul Gauguin, *Chou Quarries at Pontoise—II*, 1882, National Gallery of Canada; Vincent Van Gogh, *View of Montmartre with Quarry and Mills, Autumn*, 1886, Amsterdam, Van Gogh Museum; and the numerous paintings by Paul Cézanne of the Bibémus Quarry. Other examples of turn-of-the-century paintings of quarries include John Singer Sargent's *Bringing Down Marble from the Quarries to Carrara*, 1911, Metropolitan Museum of Art, and Paul Klee's *In the Quarry*, 1913, Klee Foundation, Bern.
5. For further information on this exhibition, see *New Topographics* (Tucson: University of Arizona, in association with the International Museum of Photography at the George Eastman House and Steidel, 2009).
6. Burtynsky's reliance upon the name "Rock of Ages" requires clarification. On one hand, it refers to a specific granite quarry, the original Rock of Ages quarry in Barre that is now inactive and largely under water. It also refers to the Rock of Ages Corporation, the commercial granite company, incorporated in 1925 and previously known as Boutwell, Milne, and Varnum, that owns all of the individual granite quarries Burtynsky photographed. Thus, in Vermont, the term "Rock of Ages" is synonymous with granite. However, Burtynsky also uses the term to refer to his Vermont quarry photographs as a unified group, including the marble quarries found in Rochester and around Rutland and Proctor that were formerly operated by the Vermont Marble Company and are now largely owned by the Vermont Quarries Corporation and the OMYA Corporation. In Vermont, the division between granite and marble quarrying is not merely due to corporate identities but also to geology: granite deposits are found to the east of the Green Mountains, and marble to the west.
7. Todd Paton, director of tourism at the Rock of Ages Corporation, who accompanied Burtynsky to various quarry sites in 1991, observes that the photograph *Rock of Ages #7*, subtitled "Active Granite Section, Wells-Lamson Quarry" in *Quarries* (2007), could not have been shot at the Wells-Lamson Quarry because it was closed for quarrying in 1986. He identifies it as the E. L. Smith Quarry. When queried by Juliette Bianco in preparation for this book, the artist agreed with Paton's observation, and the caption has been corrected here.
8. Dorothy A. Richter, "Barre Granite Quarries, Barre, Vermont," in *Geological Society of America Centennial Field Guide—Northeastern Section* (1987), 239–42. Today the E. L. Smith Quarry boasts an unusually high recovery rate of 75 percent, due to the texture and quality of its granite. Rather than being piled, the grout is dumped on "submarginal" land and backfilled. Paton, *The Rock of Ages Story*, 29.
9. Kisa Lala, "Interview with Edward Burtynsky: Photographing a Planet on the Edge," September 30, 2010, Spread ArtCulture blog, available at www.spreadartculture.com/ 2010/09/30/interview-with-edward-burtynsky(accessed Oct. 3, 2010).
10. "Edward Burtynsky, "Artist Statement," in *Quarries* (Göttingen: Steidl, 2007), 9.
11. Other FSA photographers also photographed Vermont quarries, including Russell Lee and Marion Post Wolcott.
12. Quoted in Lori Pauli, "Seeing the Big Picture," in *Manufactured Landscapes: The Photographs of Edward Burtynsky*, 23.
13. These pools in turn provide water for active quarry operations and comprise part of the local public water supply. See Richter, "Barre Granite Quarries, Barre, Vermont," 239.
14. Rock of Ages Corporation tour, August 22, 2011.
15. In the face of competition from China, few of the Vermont marble quarries still produce dimension stone. Instead, most are either abandoned or produce aggregate stone for road building.
16. Lala, "Interview with Edward Burtynsky."
17. Burtynsky, "Artist Statement," 9.
18. See, for example, David Nye, *American Technological Sublime* (Cambridge, Mass.: MIT Press, 1994).
19. Torosian, "The Essential Element," 49–50.
20. Edward Burtynsky, "Artist Statement," *China* (Göttingen: Steidl, 2005), 7.
21. Ibid.

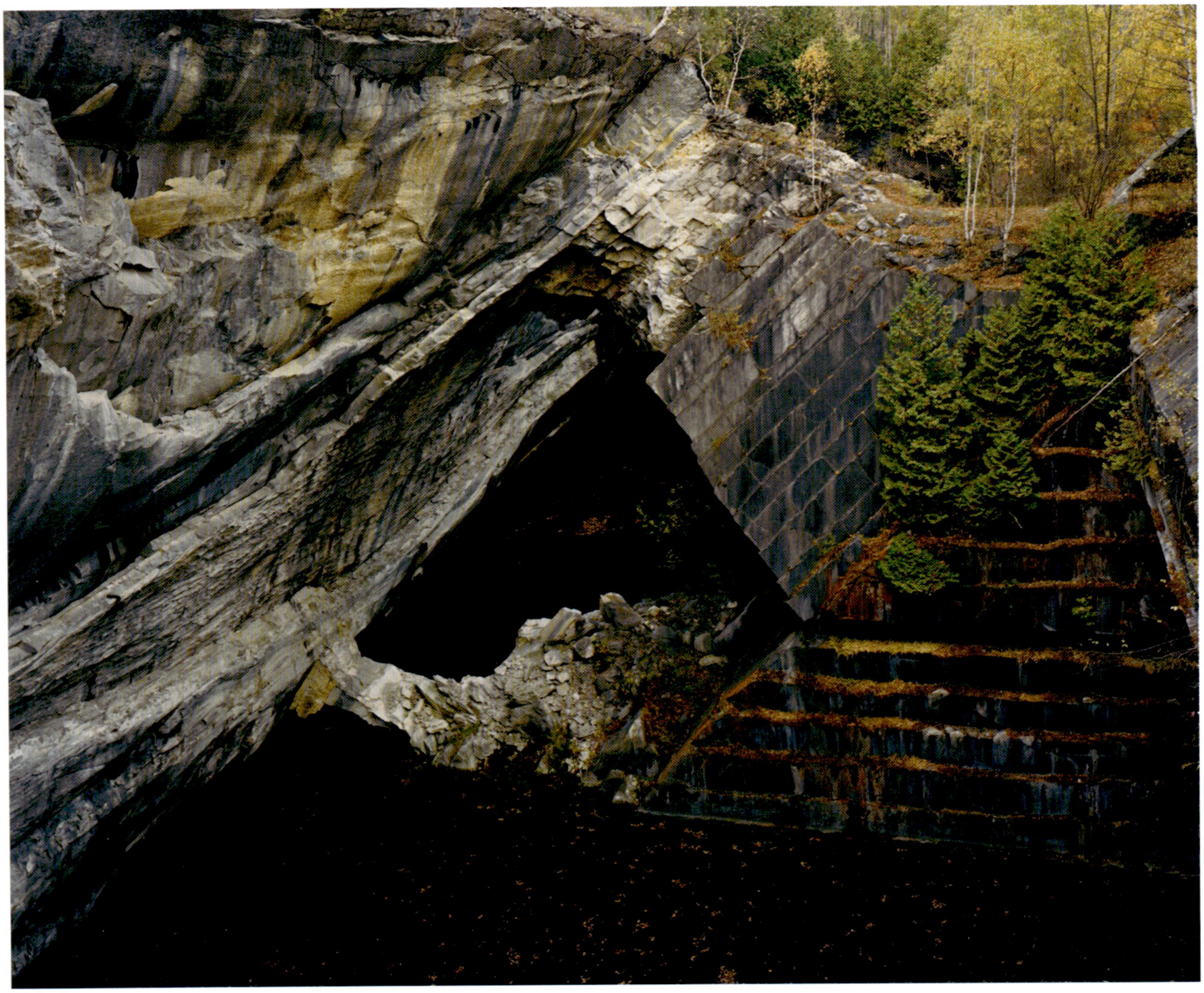

I Plates Vermont Marble Quarries

Plate 1
Edward Burtynsky
Marble Quarries #001, Rutland, Vermont, 1991
Chromogenic color print
Lent by Dennis Sughrue

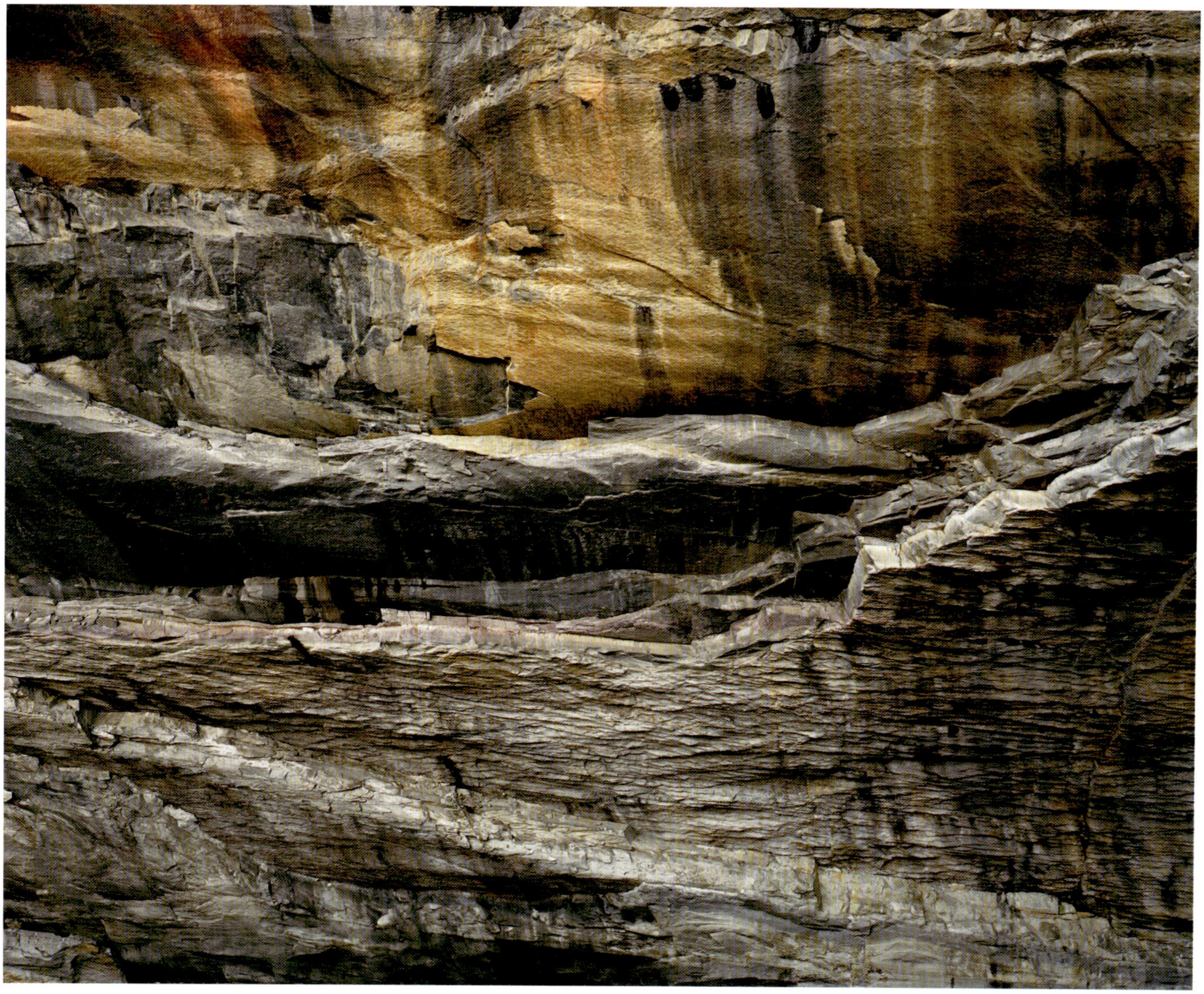

Plate 2
Edward Burtynsky
Abandoned Marble Quarry #18, Rutland, Vermont, 1991
Digital chromogenic color print
Courtesy of the artist

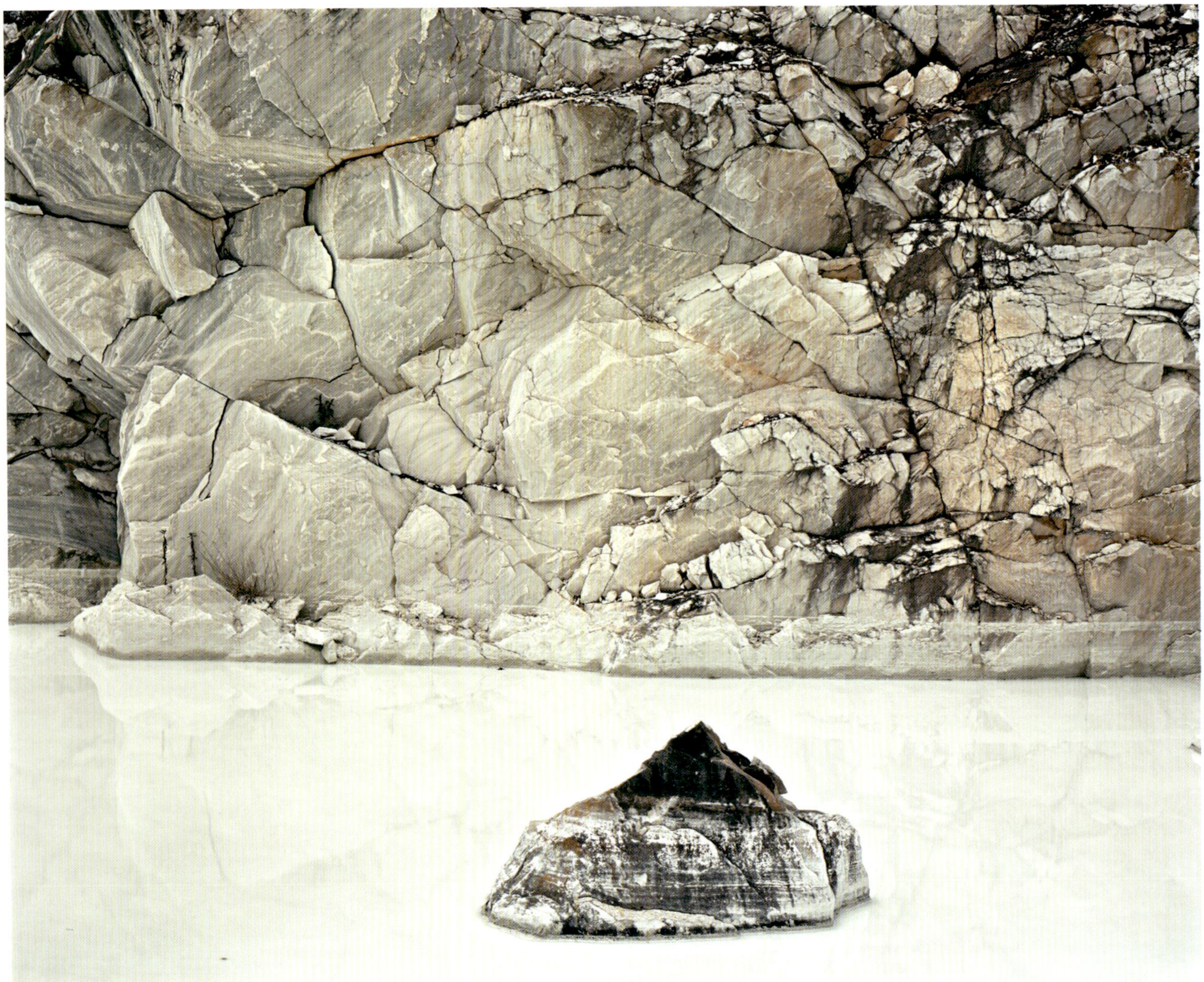

Plate 3
Edward Burtynsky
OMYA #51, Calcium Carbonate Waste Area, Proctor, Vermont, 1991
Digital chromogenic color print
Courtesy of the artist

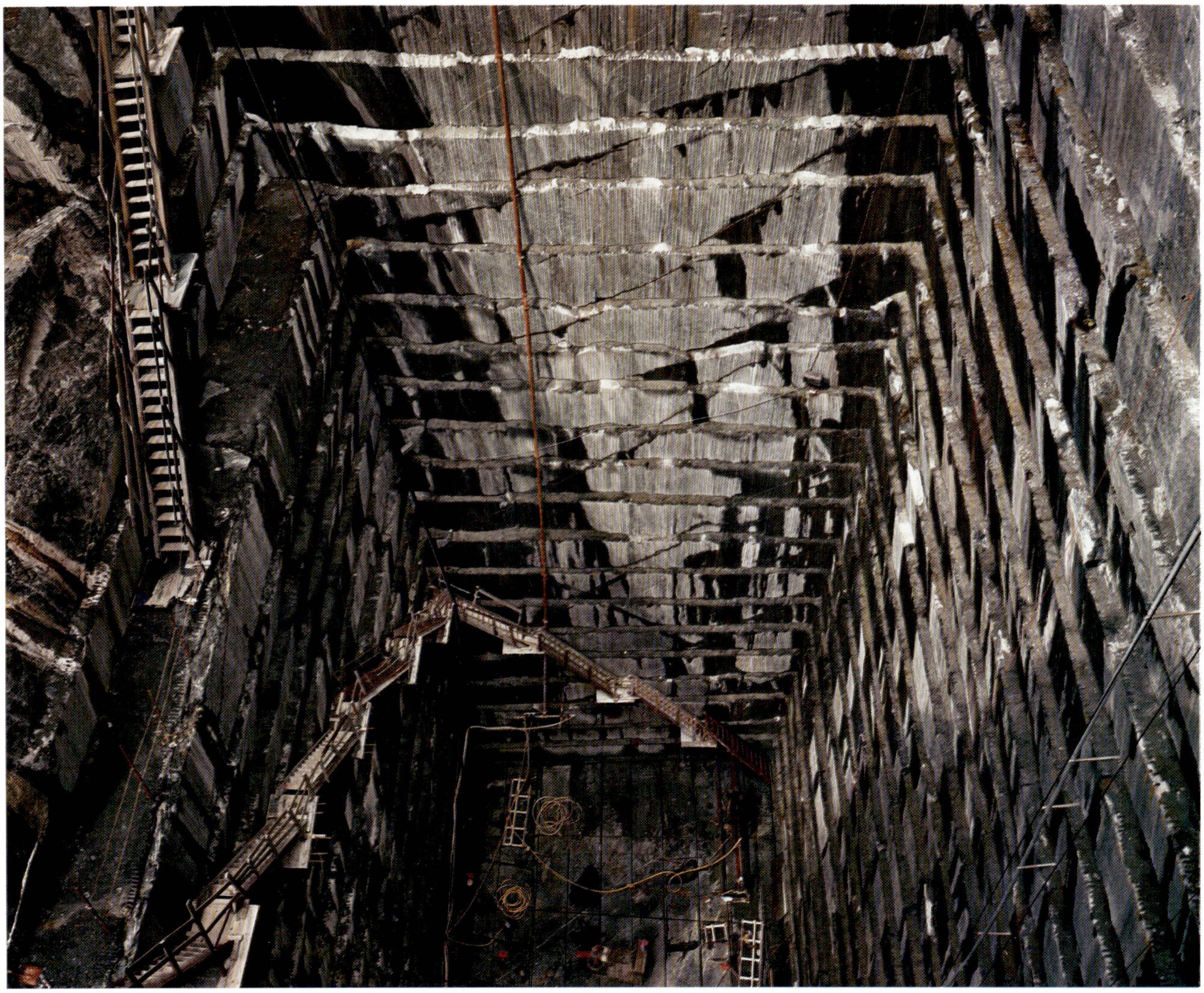

Plate 4
Edward Burtynsky
Vermont Marble Company #52, Vermont Marble Company, Rochester, Vermont, 1991
Digital chromogenic color print
Courtesy of the artist

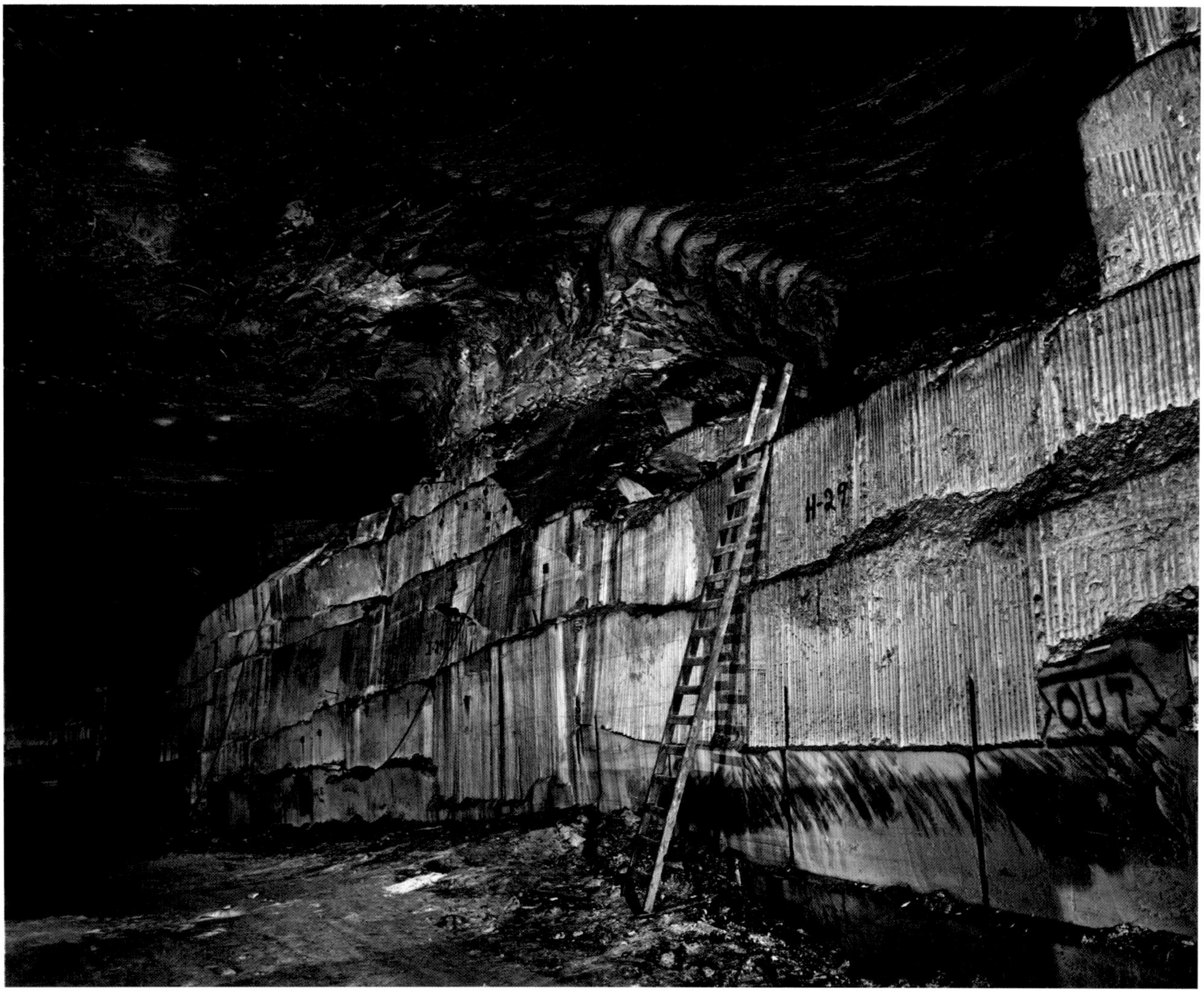

Plate 5
Edward Burtynsky
Danby Marble Quarry #1, Underground Quarry, Danby, Vermont, 1995
Digital chromogenic color print
Courtesy of the artist

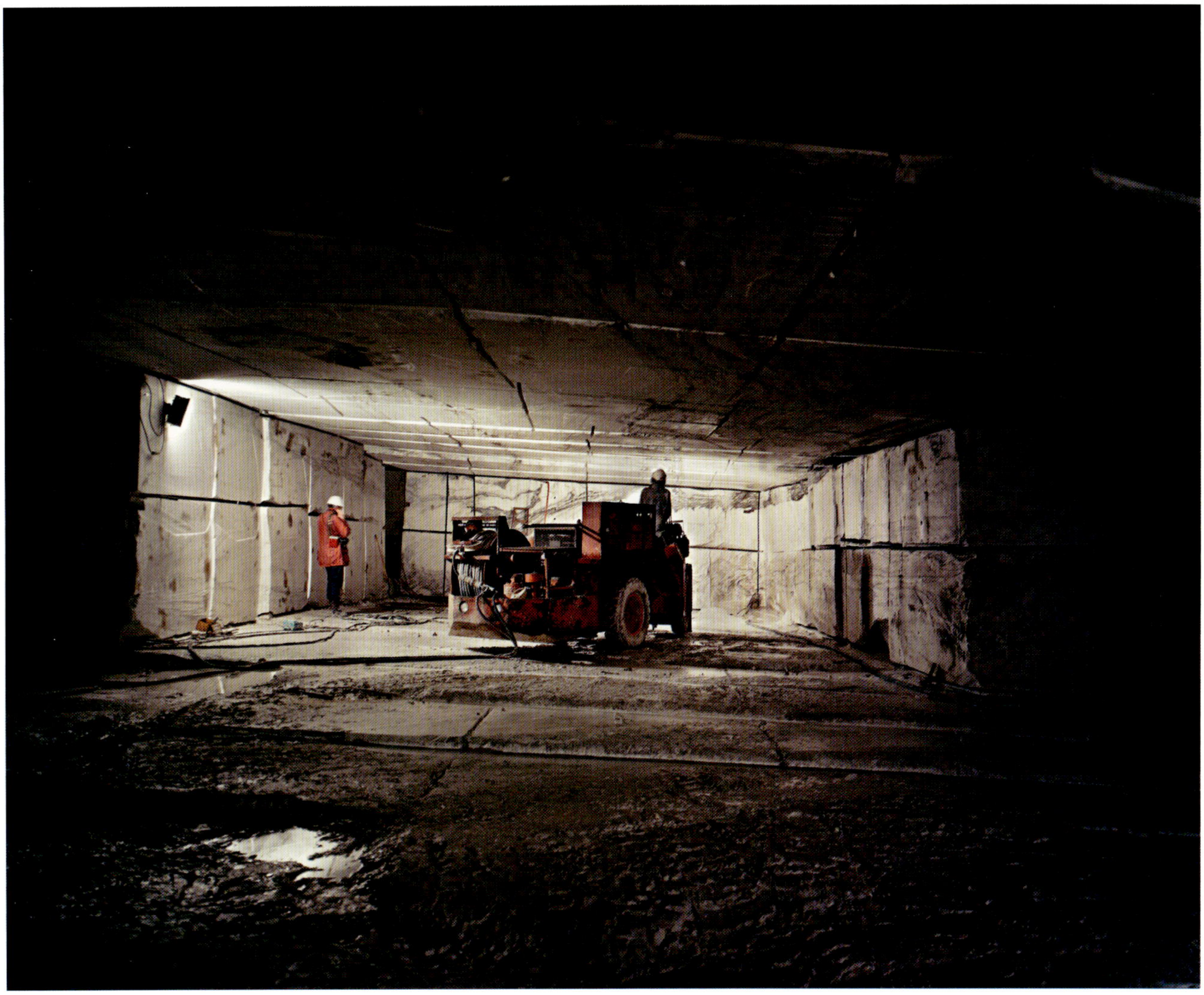

Plate 6
Edward Burtynsky
Danby Marble Quarry #2, Underground Quarry, Danby, Vermont, 1995
Digital chromogenic color print
Courtesy of the artist

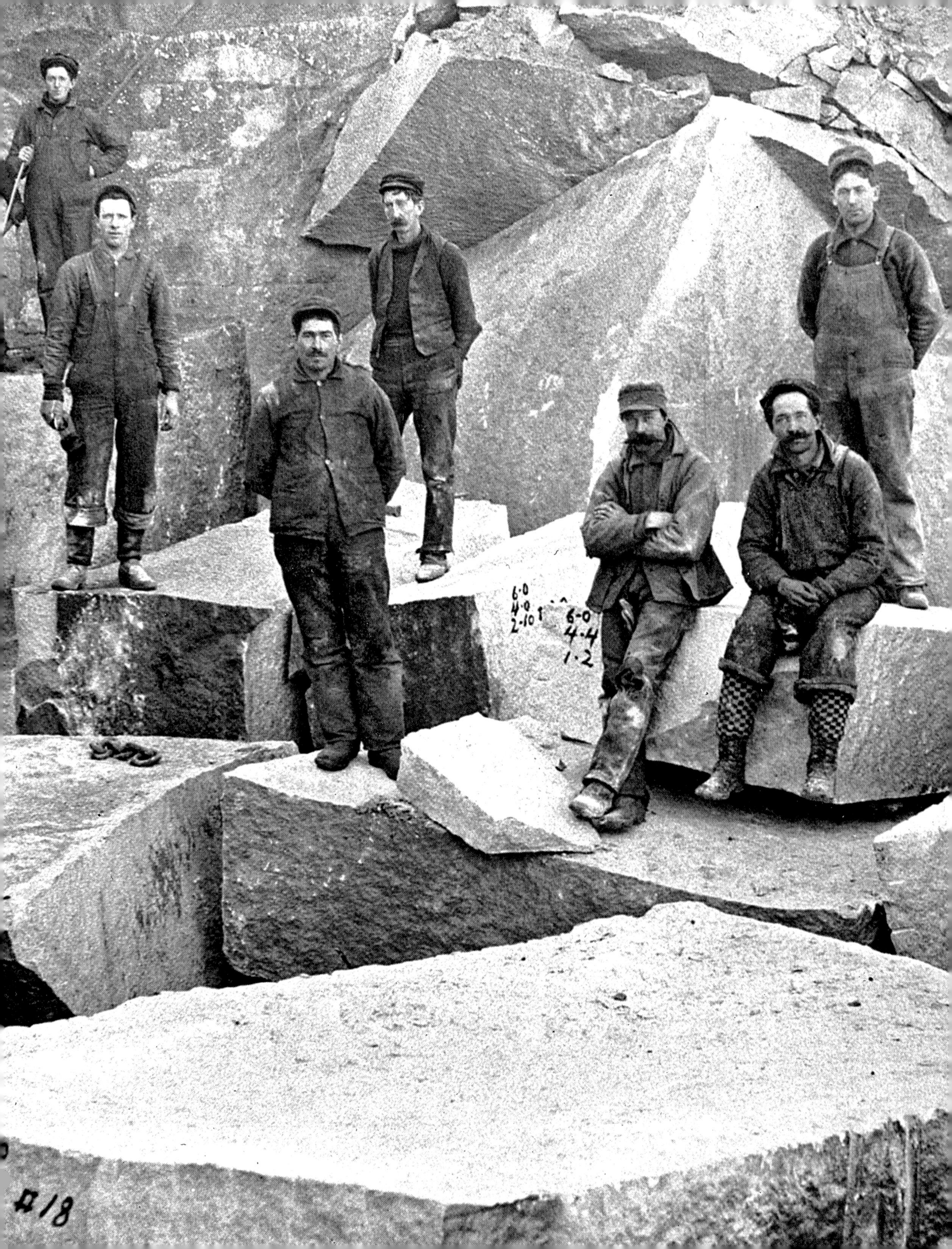
6-0
4-0
2-10
6-0
4-4
1-2
#18

3
Proctor and Barre
The Marble and Granite Quarries and the Italian Stoneworkers in Vermont, 1865–1915

Ilaria Brancoli Busdraghi

Edward Burtynsky's powerful images of the quarries of Vermont capture the grandeur of the granite and marble industries, two protagonists in Vermont's economic history straddling the nineteenth and twentieth centuries. These two stone industries have different stories, though both are equally epic. The differences start with the physical characteristics of the materials: marble is a metamorphic calcareous rock, relatively soft, homogeneous, and resistant to shattering yet not too difficult to work with, while granite is an igneous rock and extremely hard. Both can be found in Vermont in large quantities, and starting in the 1870s the industries that extracted them grew prosperous there, thanks to transportation and technological innovations as well as a national market that was more and more interested in stone.

This demand derived from specific developments in Vermont and the country at large at this time. The end of the Civil War in 1865 marked the beginning of an intense phase of urbanization; new cities were created and existing ones were rejuvenated and expanded. When it was available, urban planners favored stone, and the growth of the railway system allowed for the movement of marble and granite on an unprecedented scale. With the completion of two major railway lines in the 1870s, Vermont was forever connected to the whole of the Northeast and thence to the rest of the country. Its large quarries soon made this little state one of the nation's main providers of marble and granite. Even several decades later, an anonymous geologist carrying out a survey for the State of Vermont could wax poetic about its prospects:

> *Probably nowhere else in the known world can be found in an area equally small such valuable and extensive deposits of stones counted by mankind everywhere the most beautiful for the homes of the living and for the tombs of the dead, as may be found within the boundaries of the state of Vermont. Phidias never wrought his immortal statues from marble more beautiful than that produced within the shadows of the Green Mountains. The Parian, the Pentelicon, the Hymettos, the Carrara marbles were not superior to those of Vermont. Out of Vermont hillsides today are coming the marble and the granite that are being used in the finest temples and capitols and mausoleums in the land, as well as the most beautiful interior decorations made of stone, and there seems no danger that these deposits will fail, for the supply is practically inexhaustible.*[1]

Proctor was the center of the marble industry, and the stone was quarried from a deposit on the western side of the Appalachians.[2] The two main Vermont marble deposits run parallel to each other in the Vermont Valley, between the Green Mountains and the Taconic Range. Both contain more than one hundred varieties of marble, ranging in color from snow white to ink black.

The commercial exploitation of marble for the local market arose during the last decades of the

Photographer unknown, *Granite quarry and workers, Consolidated Granite Co., Eagan Quarry, Barre, Vermont,* date unknown. (Detail of fig. 3.3.)

eighteenth century, as the stone began to be used for decorative elements of limited dimensions. Already in the early 1800s, a number of small marble companies were active in Sutherland Falls, Rutland, and West Rutland. Doorsteps, windowsills, and mantels of Vermont marble became a staple in New England houses, and "the cemeteries of the Eastern States filled up with weeping angels, obese doves and sculpted lilies."[3] By the second half of the century, however, these companies were struggling to stay in business, and all were facing the same problems: the small scale of the applications, rudimentary extraction procedures and machinery, transportation issues—Vermont was still isolated at this time—and a lack of capital and skilled workers.

Change would come, however, in the form of technological progress and, as mentioned previously, improvements in the rail system, and eventually it would be possible to think feasibly about marble extraction, cutting, carving, and sculpting as a profitable enterprise geared toward a national market (fig. 3.1). The figure who would prove equal to this task was Redfield Proctor, a lawyer from a Yankee family in Rutland who had led the Fifteenth Brigade of volunteers from Vermont in the Battle of Gettysburg. In 1869, Proctor was nominated executor for Dorr and Meyers, a failed marble company in Sutherland Falls. His arrival in that town (renamed "Proctor" in 1882 in honor of his success) would mark the beginning of a new phase for the marble industry in Vermont. In all of the sources, Redfield Proctor is described as the sort of determined and ambitious businessman who was typical of an age that produced many magnates. David C. Gale, author of an article on Carrara and Proctor, wrote this about him in 1914: "[He] first convinced himself of the worth of American marble, then he took up the task of convincing his countrymen. He placed marble before them at every possible opportunity. He talked of it by day and dreamed of it by night. With a persistency that knows no turning back, he went into the fight—and won."[4]

Vermont marble was, indeed, of high quality, as most Vermonters were well aware. By 1880, at the conclusion of the series of acquisitions and mergers that led to the creation of the Vermont Marble Company, Proctor had become the owner of virtually all the Vermont marble and of a company that could handle the stone from quarry to finished product. This capacity gave Vermont Marble Company an advantage over most of its out-of-state competitors, and Vermont marble was used in Washington, D.C., for the Supreme Court, the Jefferson and Lincoln Memorials, and many other monuments around the country. Proctor, a small town a few miles west of Rutland, became the geographic and organizational hub of the marble business. In the span of just ten years, it grew from a small, quiet village to a multiethnic boomtown boasting the headquarters of a monopoly that did business in New York, Philadelphia, Washington, D.C., and elsewhere. By 1912, Proctor's quarries provided no less than 45 percent of all the marble extracted in the United States.[5] But in 1880, despite all of his progress, Redfield Proctor still had a problem: the lack of skilled workers to allow him to take on the largest projects. He needed cutters, carvers, and sculptors who, on paper and in stone, could realize much more elaborate and artistic models than those with which he had thus far contented himself.

Those interested in Vermont's sister stone shared his woes. The Green Mountains divide Proctor from Barre, which, to this day, calls itself "the granite capital of the world." Its granite area was about four miles southeast of the city, with most of the quarries operating in a territory three miles long and one mile wide.[6] Already in the eighteenth century the native Abenaki people were making use of the stone from the deposits in Barre, and early European settlers promptly used it for mills, window and door frames, doorsteps, mantels, and the like. In 1815 the first proper quarry was opened on Millstone Hill, overlooking the town of Barre. The granite from these quarries is of high quality; in particular, it contains very little iron, which leaves it free from stains even when it is exposed to the elements. Between 1833 and 1837, stone extracted from Millstone Hill was used for the Vermont State House in Montpelier. At that point, the blocks of granite were still moved by teams of oxen at the speed of about four miles per *week*. Transportation of the larger blocks would even have to wait for winter, when sleighs would permit faster transport (fig. 3.2). Moving the material was such a challenging and risky endeavor that, at the time of the building of the Vermont State House, a government document made this prophecy: "This is the last structure that will ever

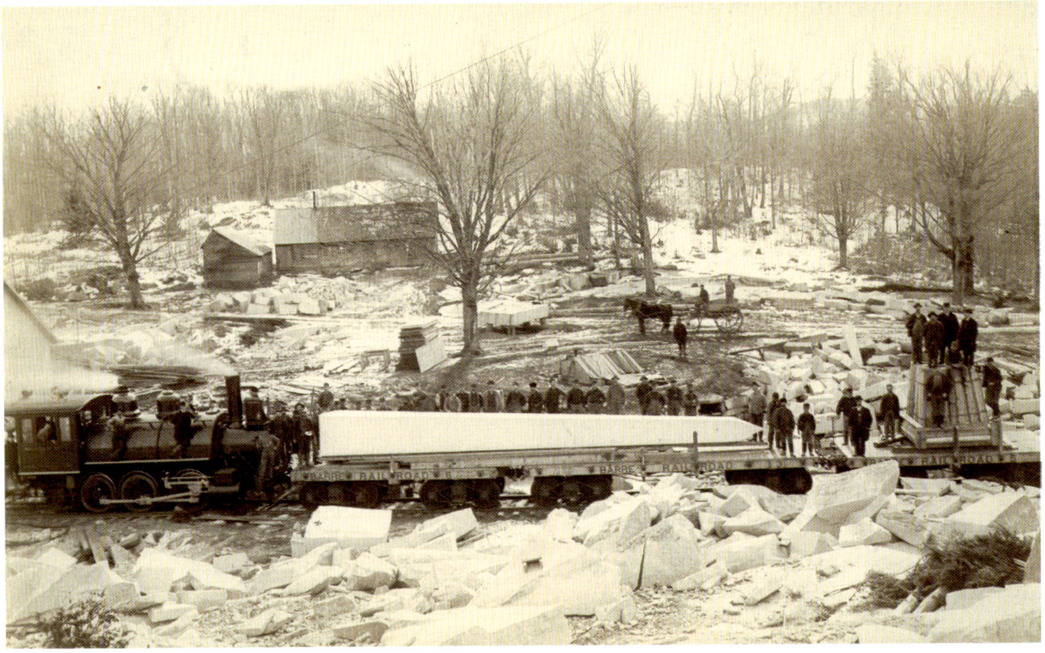

Figure 3.1
Photographic Studio of F. W. Sherburne,
Barre, Vermont, *Barre Railroad,* about 1890.

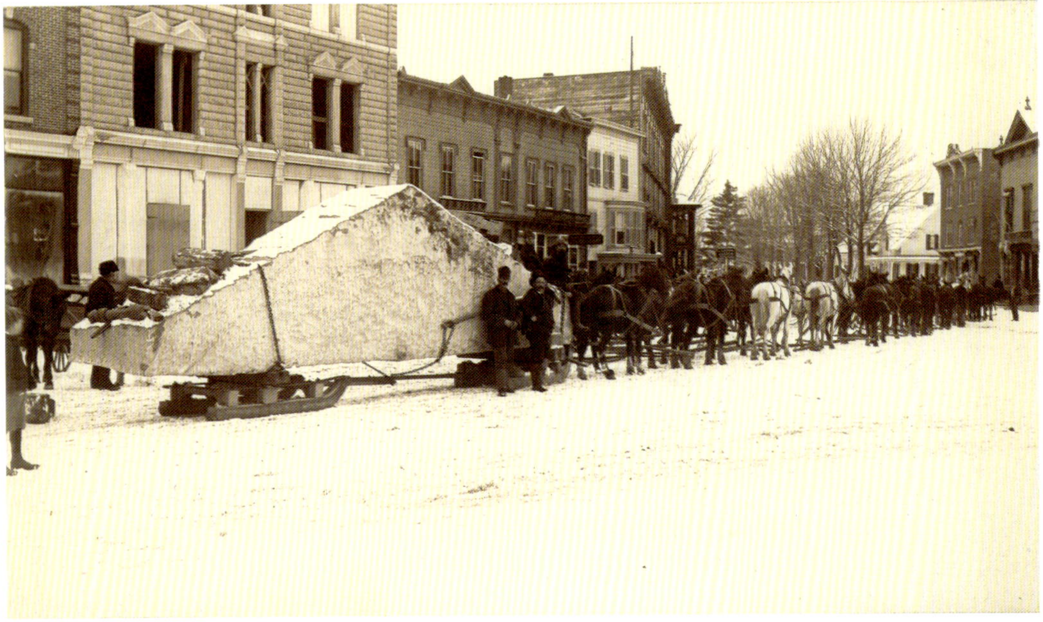

Figure 3.2
Photographic Studio of F. W. Sherburne,
Barre, Vermont, *In Barre; part of Stanford Mausoleum; weight of stone 54 tons,* 1888.

be built of Barre granite."[7] More commissions followed nevertheless, but as in Proctor and Rutland, it was not until the completion of the railway in 1875 that the transportation problem was resolved. At the same time, Barre was made more easily accessible to the skilled workers that the granite industry, like the marble industry, sorely lacked. The need for experienced stoneworkers brought many immigrants to Vermont: Swedes, Scots, Poles, Spaniards, and Italians, among others, who settled near the quarries and practiced their trades of mining, cutting, carving, and sculpting the stone. Their presence transformed the villages of Proctor and Barre into multiethnic towns, but the differences in the experiences offered by the respective industries of marble and granite were profound.

This discrepancy was especially visible among the Italians. They arrived from the same nation, and with similar backgrounds, but their experiences in Proctor and Barre took dissimilar paths; those towns, though involved in the same kind of trade and business, differed from each other in substantial ways. The Italians who migrated to Vermont between 1880 and 1915 were for the most part from Carrara, in northern Tuscany, and the pre-Alpine valleys of Piedmont and Lombardy. They were decidedly not "professionally very close to a pure supplier of muscular strength," as Ercole Sori describes the bulk of the Italian participants in the "Great Migration."[8] They were, instead, artisans and artists—highly skilled stoneworkers whose expertise had been shared across generations. When they brought their talents to Vermont, they also brought their strong values and political ideals, a defined identity that assisted their integration into American society. To many of these people, migration was not a new experience; in the pre-Alpine valleys of Piedmont and Lombardy, migration had represented a solution to economic hardship for centuries. Migratory flows established already in the 1200s were still active at the end of the 1800s. Skilled workers seasonally migrated south, into the Italian peninsula, or to the Mediterranean basin in general, or even to central and northern Europe. These were the men that historian Raul Merzario has called "the aristocracy of emigration."[9] Alongside silk workers, animal tamers, and chimneysweepers, there were also many stoneworkers. In those valleys the granite and marble quarries provided alternatives to the meager economy that the rough, hard soil of their native hills and mountains allowed.

This long experience of migration created a sprawling web of relationships based upon a shared trade and sometimes even a shared extended family.[10] In the last decades of the nineteenth century and the early decades of the twentieth, these relationships and general know-how would be invaluable to those who decided to migrate, this time overseas. As Rudolph Vecoli wrote, "These forerunners, seeking markets, customers and work, established the itineraries, the mental maps" for those who left after them.[11] The Tuscans from the area of Carrara in particular had been extracting marble for nearly two millennia, and specialized quarrying, cutting, carving, and sculpting skills had been passed on from father to son (plates 7–10). They were the less adept world travelers, however. In spite of the risks connected to quarrying and working marble, and the fact that the technology they used was old and dangerous, the stoneworkers in this region were more rooted in the territory and traditionally less likely to migrate.

For all these stoneworkers, those from Piedmont and Lombardy as well as the Tuscans, the pervasive confidence in their skills produced a collective identity that had been buttressed since the Middle Ages by their membership in guilds and, later, unions. These expressions of solidarity offered a safety net, from both an economic and a cultural point of view. Their membership in labor organizations also meant that these workers, unlike the vast majority of Italians, were actively involved in politics; Carrara, for example, was one of the most important centers of Italian anarchism, even in the marble quarries and sheds: "Compared with other parts of Italy, Carrara marble workers developed an early interest in politics, and unlike the majority of Italians, had definite views about the unification of Italy . . . They hated the oppressive governments."[12] The same was true in Piedmont and Lombardy, where the stoneworkers were for the most part socialists.

One of the manifestations and outcomes of political activism was the publication of newspapers that in turn supplied a wealth of information about the emigration experience to colleagues at the time and scholars in later decades. These pages deal with every aspect of the emigration experience and life in

America: how to travel, where to find a job, how to buy a house. The combination of identity and solidarity would be an essential element during the migration and settlement experience. Redfield Proctor had business relations with Walton, Goody, and Cripps, a British company with offices in Carrara, and through its help he convinced five Italian stoneworkers to teach the Americans their skills. They migrated to Vermont with the idea of working there for a few years, taking advantage of the better pay, and then returning to Italy, so none of them took his family. This would change within a few years: in 1885 the wife of Oreste Bertagna arrived in Proctor, a sign that Bertagna's experience in Vermont, at least, was no longer thought to be temporary. In 1937, Roberto Burattini, author of a study titled *Italians and Italo-Americans in Vermont*, wrote: "The majority arrived with the intention of staying for a short period, but when they found out that their house was good and comfortable, they made it permanent."[13]

In Proctor and also Rutland, these skilled Italian immigrants worked almost exclusively in workshops where the marble was cut, carved, and polished. They made the preparatory drawings, sculpted the marble, and taught their artistic concepts and manual skills to other carvers. Only rarely did they work in the quarries extracting the stone; this activity, much more taxing and dangerous, was generally performed by the immigrants from Poland, Ireland, and Sweden who had started to settle in that area after the Italians came.

As in so many other places in the United States and elsewhere, the Italians tended to settle in the same cluster of streets, where people of the same regions (if not towns) would surround them, speaking the same dialect. The familiarity of the language and social landscape, as well as the scenery, eased the transition of these men and women into their new homes. "Why did you come all the way here from Italy, to bury yourself in a small village in the mountains, far away from big cities and lacking almost all the modern amenities?" asked Roberto Burattini to an Italian cutter. The man smiled and replied: "But in Italy we used to live in a small village just like this, so we live here as if we were at home."[14]

Another way these people reaffirmed their identity and found solutions to the many difficulties they faced was by duplicating Italian organizations in their new lives. The most important was the Italian Aid Society, founded in Proctor in 1894. This association provided its members with a support system and a venue through which Italians, at first exclusively from the province of Massa Carrara, could meet and organize activities. A food committee was created to manage a shop that sold items—olive oil, pasta, polenta, *baccalà*, and dried beans—they could not easily find in existing stores. The role that organizations such as this one played can hardly be overestimated. Geno Pisanelli, who arrived in Rutland in the winter of 1909 on a day so cold that all the roads were covered with ice and he had to crawl part of the way, said of the aid society: "In it I found a support that reminded me of my house and that helped me keep soul and body together while I was getting used to life in a strange and cold place."[15]

In Proctor, however, the space for political involvement felt severely limited to men who were used to expressing their political ideals. This was due to the ways in which Redfield Proctor had built his successful business: he not only made money but also controlled the lives of his employees. The company rented or sold houses to its workers, ran a cooperative store and a hospital, and organized holiday festivities for all. Harmony in the workplace and in town was what Redfield sought more than anything, at the expense of any involvement in or even discussion of politics. In 1914, David Gale wrote: "Whenever it happens that an employee becomes addicted to rowdyism and persists in an attempt to set up a lower standard of living, he is told that his services are no longer needed. And the same rule is applied to all undesirable influences. The little village has never refused to open its gates to anything that would contribute to its social betterment, but it has persistently frowned upon all other agencies."[16]

This had important consequences for those workers who refused to renounce the ability to express their political ideals, an essential part of who they were. Asked to go elsewhere, many of them opted for Barre, on the eastern side of the Green Mountains. The story of the Italians in Barre's granite industry differs from the story of the Italians in Proctor and Rutland, both because their arrival was more roundabout and because Barre had a character all its own. Recruitment

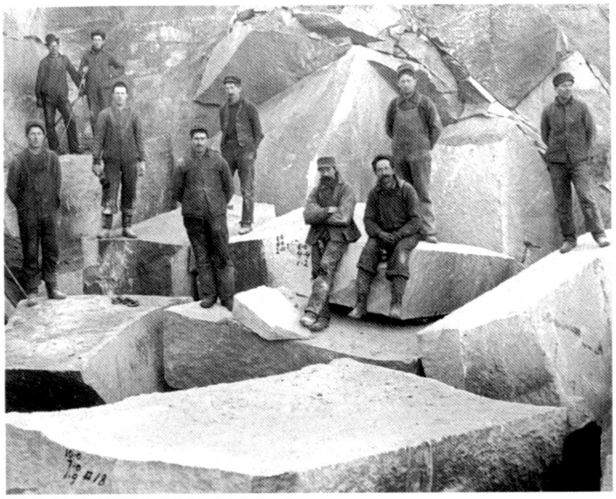

Figure 3.3
Photographer unknown, *Granite quarry and workers, Consolidated Granite Co., Eagan Quarry, Barre, Vermont,* date unknown.

of this equally specialized workforce happened in several ways: publications such as the *Granite Cutters Journal* listed the available jobs, information traveled across the ocean by way of letters and articles, and the American Emigration Company employed workers on behalf of American enterprises. The first outsiders to migrate to Barre were, in 1880, Scots mostly from the region of Aberdeen, in the north of that nation, where the granite business was going through a serious crisis. Some were cutters and sculptors, but for the most part they were skilled quarry men, and they arrived in Vermont seeking better pay (figs. 3.3 and 3.4).[17] Aberdeen had a strong unionist tradition, even relative to the rest of the United Kingdom; once the Scots arrived in Vermont, then, they immediately became proactive members of the Granite Cutters National Union. Barre's section of the GCNU was organized in 1886, and by 1900, when the community's Scots and Italians had combined forces, it had become the largest section in the United States, with more than one thousand members.[18] The early presence of the Scots shaped the whole organization of Barre's granite business: instead of one overarching powerbroker, a myriad of small sheds worked alongside one another—in the 1890s there were over one hundred—and rewarded nothing so highly as one's individual skill with the stone.

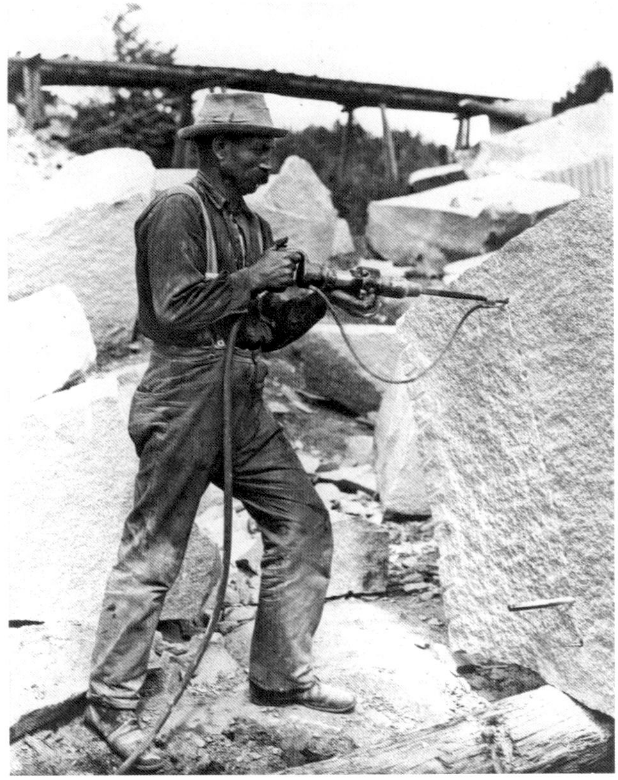

Figure 3.4
Photographer unknown, *Granite quarrier using pneumatic drill,* about 1900.

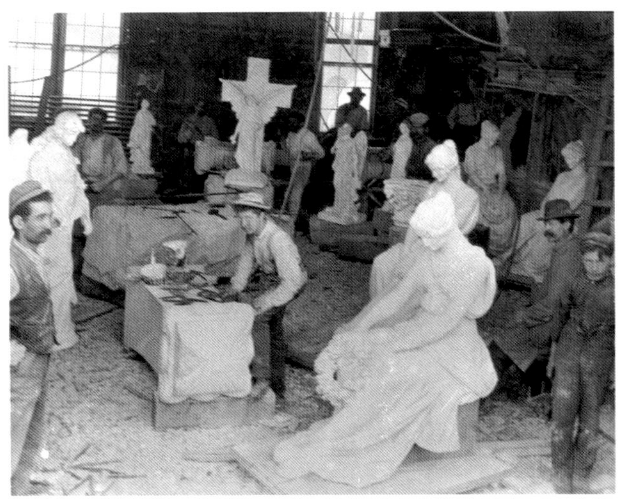

Figure 3.5
Photographer unknown, *Interior of the Corti and Novelli Shed (Corti is at the far left; Novelli is at the right, seated),* about 1902.

Compared to Proctor and Rutland, then, Barre set a very different stage for the Italians who a few years later followed the Scots, either directly from Italy or after working in quarry towns in other states, or, of course, in Proctor (fig. 3.5). Once the Proctor stone-worker had modified his skills in relation to the new material, he found Barre to be a more diversified and open workplace that in fact encouraged the expression of one's political views and participation in unions. As Paul Demers wrote: "[In Barre] labor was not a commodity; labor was a way of life."[19] In those years the town grew quickly: a village of 2,000 people through about 1880, Barre grew to 6,800 in 1890, 8,500 in 1900, and more than 11,000 in 1910. Between 1881 and 1893, 625 houses were built, an average of one house per week for twelve years.[20] An Italian neighborhood arose in the north part of the city, the so-called North End. As Maila Pentucci wrote, "The massive displacements of these skilled workers, attracted also by the high wages, their compact elite identity, specialized and better paid within the migratory phenomenon, favored the creation . . . of an 'arrival community' in Barre that replicated in all the one they had left."[21] As in Proctor and Rutland, rice, polenta, pasta, homemade wine, and grappa were staples on the tables of the Italians. Vegetables and grapes grew in the small gardens in front of the houses, and mushrooms were put to dry in the sun. The ingredients to prepare

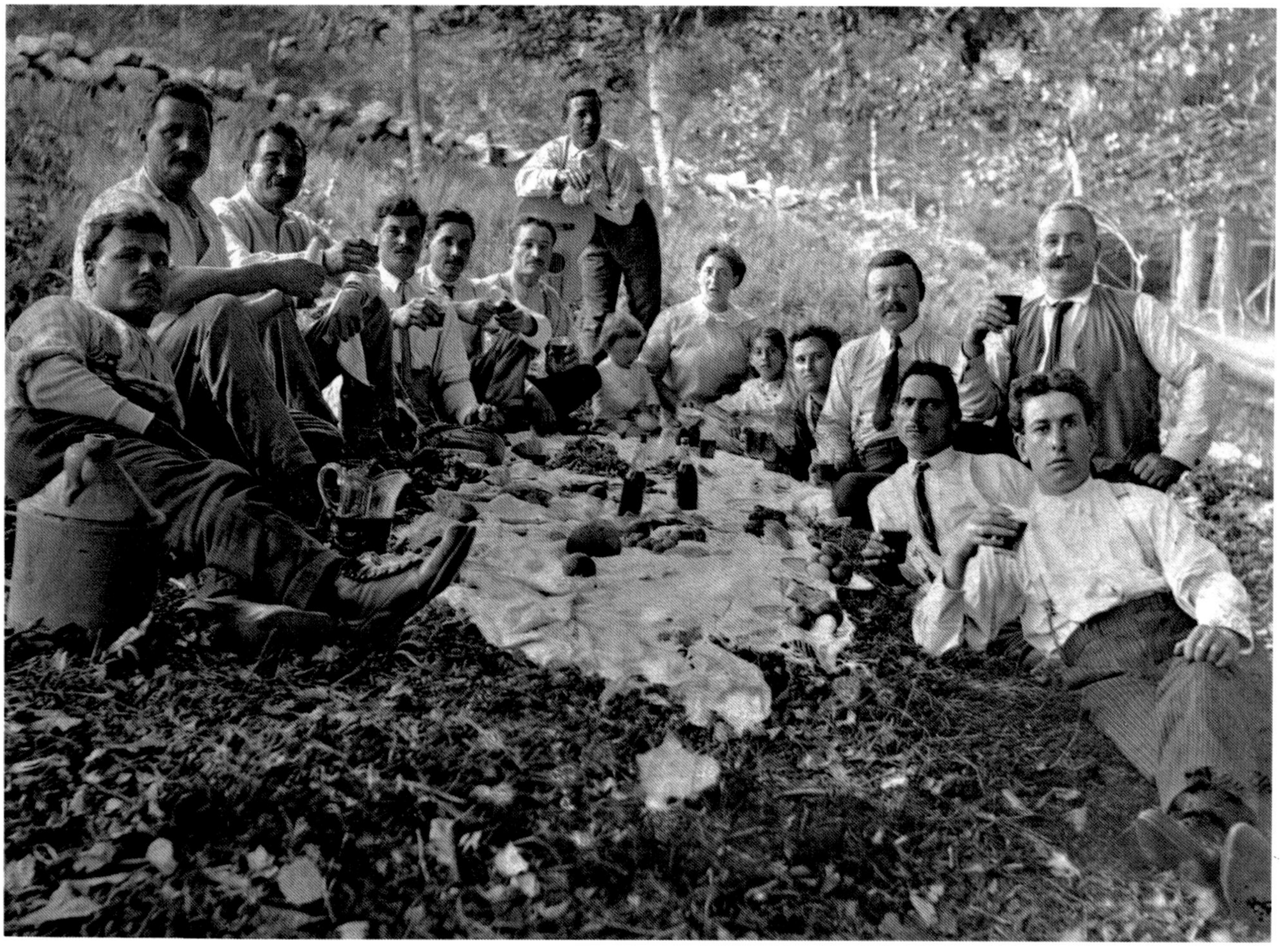

Figure 3.6
Photographer unknown, *Picnic in a field above the North End in Barre, Vermont*, date unknown.

Italian dishes could also be bought in the stores that other Italians immigrants, often from the same regions as the stoneworkers, had opened in Barre. On Sundays there were big picnics with food, games, music, songs, and dances (fig. 3.6).

In Barre, too, the Italian immigrants created various associations for mutual help and recreation, such as the Mutual Aid Society, the Italian Pleasure Club, the Columbus Band, and the Italian Athletic Club (figs. 3.7 and 3.8). The place that more than any other came to represent them and their ideals was the Socialist Labor Hall. Built by the Italian stoneworkers, it opened on November 28, 1900, and to this day it stands on Granite Street. Seven hundred Italians participated in its opening ceremony, and Camillo Cianfarra, editor of *Il Proletario*, a socialist publication from New York, gave a speech titled "What Is Socialism?" In a celebratory article about the event, he wrote: "What will at first seem impossible is a *fait accompli*, as the Hall stands now on Granite Street, a superb and masterly synthesis of the collective effort of workers united and guided by the ray of an idea like ours. To many, perhaps, this Hall will mean nothing, for many others it will be the subject of jokes; but for me, when I saw it for the first time, it spoke a language full of wisdom" (figs. 3.9).[22]

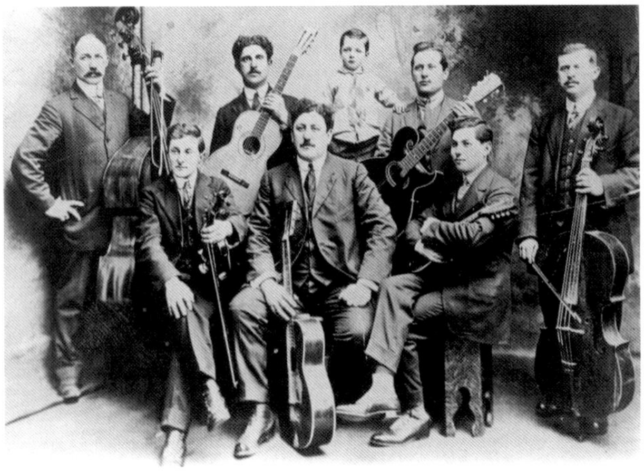

Figure 3.7
Photographer unknown, *Veronese band,* 1916.

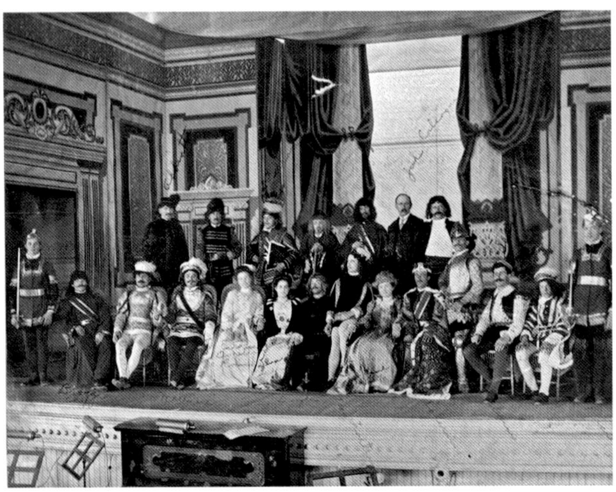

Figure 3.8
Photographer unknown, Il Fornaretto di Venezia *(The baker of Venice) cast from the production by La Vecchia Filodramatica Society at the Barre Opera House,* 1902.

The hall became the center of the Italian community, a place for social gatherings and political activities. In the big room on the ground floor there were weekly balls, for the most part organized to raise funds for the families of the many stonecutters who died of silicosis. Social activities of all kinds took place there, including weddings, card games, and English lessons. A cooperative store was opened in the building, and a bakery with a brick oven appeared behind it.[23] It was the meeting place of the Granite Cutters National Union and the stage for heated debates between the socialists and the anarchists. It also served as a refuge of sorts. At the beginning of 1912, for example, the textile workers of Lawrence, Massachusetts, went on a strike to protest a pay reduction and working conditions. In spite of intimidations by employers and authorities, the strike continued for weeks, and the situation for the families of the strikers became extremely hard. In a gathering at the Socialist Labor Hall, the Italian stoneworkers decided to pay for some of the "Lawrence children" to come to Barre (fig. 3.10). When the thirty-five children arrived, they were brought to the hall, and in a large room with a portrait of Karl Marx on either side of a stage, a big banquet took place. All of the children were photographed, and their photos were then sold to raise funds for the strikers. The activities held at the hall thus display the ideals of community and solidarity shared by the Italian immigrants, and their ability to reach beyond their new community in Barre to the larger American society of which they were now a part.

One of the most important celebrations that took place every year in the Labor Hall was the First of May, *il Primo Maggio*, which became known in Barre as the Italian Labor Day. Debates, fairs, balls, and

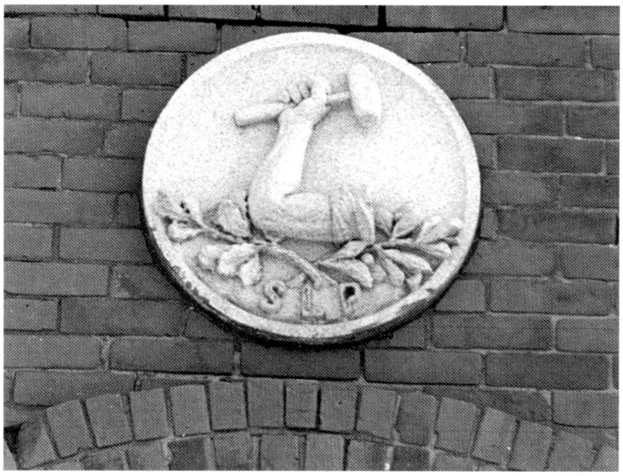

Figure 3.9
Photographer unknown, *Socialist Labor Party symbol above the door of the Socialist Labor Hall, Barre, Vermont,* date unknown.

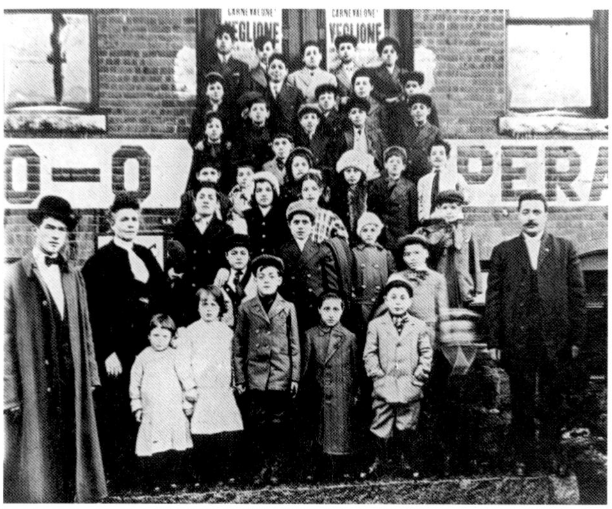

Figure 3.10
Photographer unknown, *Children of Striking Lawrence, Massachusetts,* February 15, 1912.

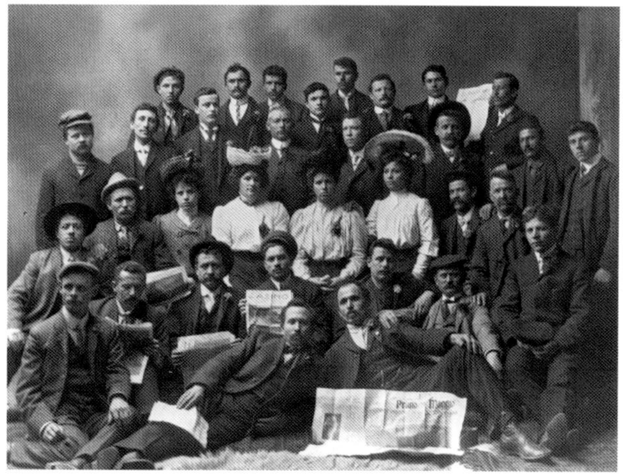

Figure 3.11
Photographer unknown, *Il Primo Maggio (The first of March—Italian Labor Day), Barre, Vermont,* 1906.

plays were organized, and the celebrations continued in the streets of the town with a parade accompanied by a band (fig. 3.11). This event underscores once again the difference between Proctor and Rutland, on one hand, and Barre, on the other: on the first of May 1909, Proctor officials inaugurated its Recreation Club, an association dedicated solely to entertainment, as its bylaws indicate: "The goals of the Recreation Club are two-fold: recreation and entertainment. The discussion of Work, Politics and any other topic that might be offensive for other members is strictly prohibited." As Anthony Pisanelli and Umberto Trinci, sons of two Italian stoneworkers from Carrara, observed: "[In Proctor, the Italians] were less involved in politics than the Italians in Barre. They were great family people. This is what mattered to them."[24]

While these various organizations and activities did not eliminate the hardship of emigration, they did allow the Italian stoneworkers, through their own initiatives, to take a proactive role in Barre's life. As elsewhere, the presence of the Italian stoneworkers is an important part of Vermont's history in the years between 1880 and 1915. In a period when the stone industry was booming, their centuries-old expertise was highly sought after, and the contribution they made to the success of the stone industry is to this day a source of pride and recognition. "As a thread in

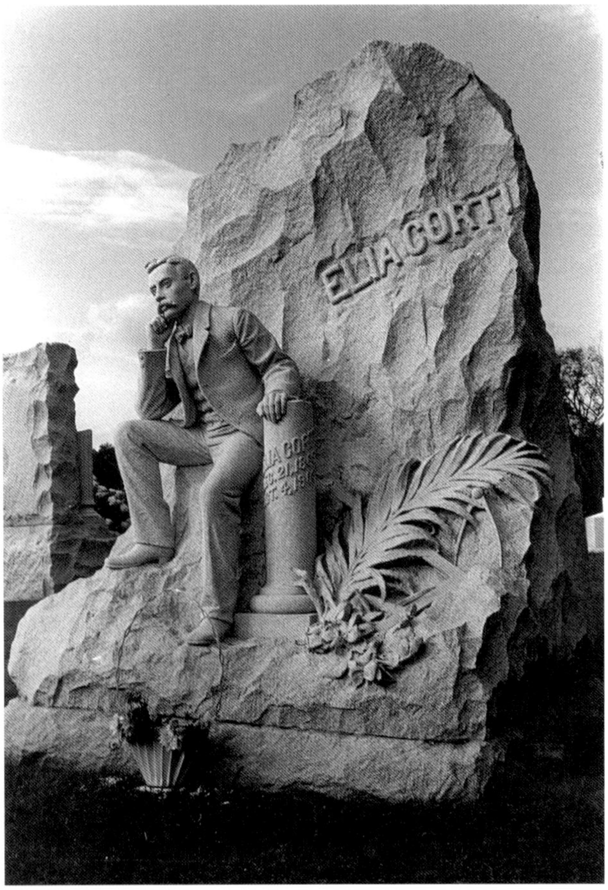

Figure 3.12
Photographer unknown, *Elia Corti Monument, Hope Cemetery, Barre, Vermont,* carved by William Corti and John Comi, about 1903.

the fabric of our society, the Italian heritage in many ways binds our community," said former Vermont senator Jim Jeffords in 1994, commemorating, in the U.S. Senate, the hundredth anniversary of the Italian Aid Society. This society, the Socialist Labor Hall, the Opera House, and Hope Cemetery in Barre (fig. 3.12 and p. viii) with its evocative funerary monuments—and, of course, the great marble and granite quarries featured in Edward Burtynsky's photographs—these are elements that tell a very tangible story of the Italian presence in Vermont. Through their work and lives, the Italian immigrant stoneworkers contributed to Vermont's distinctive character. The deep interconnection between specialized skills, ethnic identity, political beliefs, and self-worth shaped their experience of integration while enriching the society in which they came to live.

Notes

1. Vermont Bureau of Publicity, *Industrial Vermont: The Mineral, Manufacturing, and Water Power Resources of the Green Mountain State* (Montpelier: Capital City Press, 1914), 4.
2. Kevin Dooley and Norman Herz, "Provenance and Determination of Early American Marbles," *OSMOSA* (1994): 245. In this essay Proctor is often singled out because it was the starting point of the marble industry in Vermont. Proctor, however, shared its marble history and industrial development with Rutland, just a few miles away.
3. Robert Gilmore, "The Vermont Marble Company: An Entrepreneurial Study, 1869–1939," *New England Social Studies Bulletin* 14, no. 2 (1956): 234.
4. David C. Gale, "A Marble City and a Marble Village," *New England Magazine* (June 1914): 199.
5. Ibid., 18.
6. Vermont Bureau of Publicity, *Industrial Vermont: The Mineral, Manufacturing, and Water Power Resources of the Green Mountain State* (Montpelier: Capital City Press, 1914), 7. Marble and granite were not the only stones quarried in Vermont. In this same period, the slate industry, based in Rutland, was also very active and prosperous. In 1912 around one hundred quarries extracted slate for sale to clients as far away as England and South Africa; only Pennsylvania outweighed Vermont in slate production, and the two states combined met almost 90 percent of the world's demand for slate.
7. As reported in *Industrial Vermont*, 7.
8. Ercole Sori, *L'emigrazione italiana dall'Unità alla seconda guerra mondiale* (Bologna: Il Mulino, 1979), 32. Translations are by the author of this essay, unless otherwise noted.
9. Raul Merzario, *Adamocrazia. Famiglie di emigranti in una regione alpina* (Bologna: Il Mulino, 2000), 33.
10. See Patrizia Audenino, "The Paths of the Trade: Italian Stonemasons in the United States," *International Migration Review* 20, no. 4 (Winter 1986).
11. Rudolph J. Vecoli, "The Italian Diaspora, 1876–1976" in *The Cambridge Survey of World Migration,* ed. Robin Cohen (Cambridge: Cambridge University Press, 1995), 114–15.
12. Luciana and Tiziano Mannoni, *Marble: The History of a Culture* (Genova: Sagep, 1984), 4.
13. Roberto Burattini, "Italians and Italo-Americans in Vermont" (unpublished manuscript, Vermont Historical Society, Montpelier, 1937), 3.
14. Ibid., 10.

15. Geno Pisanelli, quoted by his son Anthony in "100 Years Later, Italians Celebrate Their Vermont Roots," *Boston Sunday Globe*, June 26, 1994.
16. Gale, "A Marble City and a Marble Village," 200.
17. Paul Demers, "Labor and the Social Relations of the Granite Industry in Barre" (BA thesis, Goddard College, 1974), 3. In Scotland the granite industry went through periods of recessions in 1878–79, 1884–87, and 1892–95. At times the unemployment rate reached 50 percent.
18. Karen Lane, "Old Labor Hall, Barre, Vermont: Preserving Working Class Icon," *Labor's Heritage* (spring/summer 1999): 50.
19. Demers, "Labor and the Social Relations of the Granite Industry in Barre," 15.
20. *Barre in Retrospect II, 1776–1995* (Barre: The Friends of the Aldrich Public Library, 1995), 7.
21. Maila Pentucci, "Realtà e voci politiche nel contesto del fenomeno migratorio: gli anarchici di Barre, Vermont, e 'Cronaca Sovversiva,'" in *Emigrazione e territorio: tra bisogno e ideale*, ed. Carlo Brusa and Roberto Ghiringhelli (Varese: Edizioni Lativa, 1995), 200.
22. Camillo Cianfarra, "Da New York a Barre (Appunti di viaggio)," *Il Proletario* 5, no. 49 (December 8, 1900).
23. In 1998 the Socialist Labor Hall was recognized as a national landmark by the National Register of Historical Places; it is one of the few monuments connected to labor history to receive this recognition. The great hall is now open and used by the whole community, and a restoration is underway, funded mostly by local fundraising and volunteer work, thus replicating the "philosophy of community" that had been a central belief of its builders in 1900. See Lane, "Old Labor Hall, Barre, Vermont."
24. Anthony Pisanelli and Umberto Trinci, interviewed by Yvonne Daley for her article "95 Years for Italian Aid Group," *Rutland Daily Herald*, February 11, 1989.

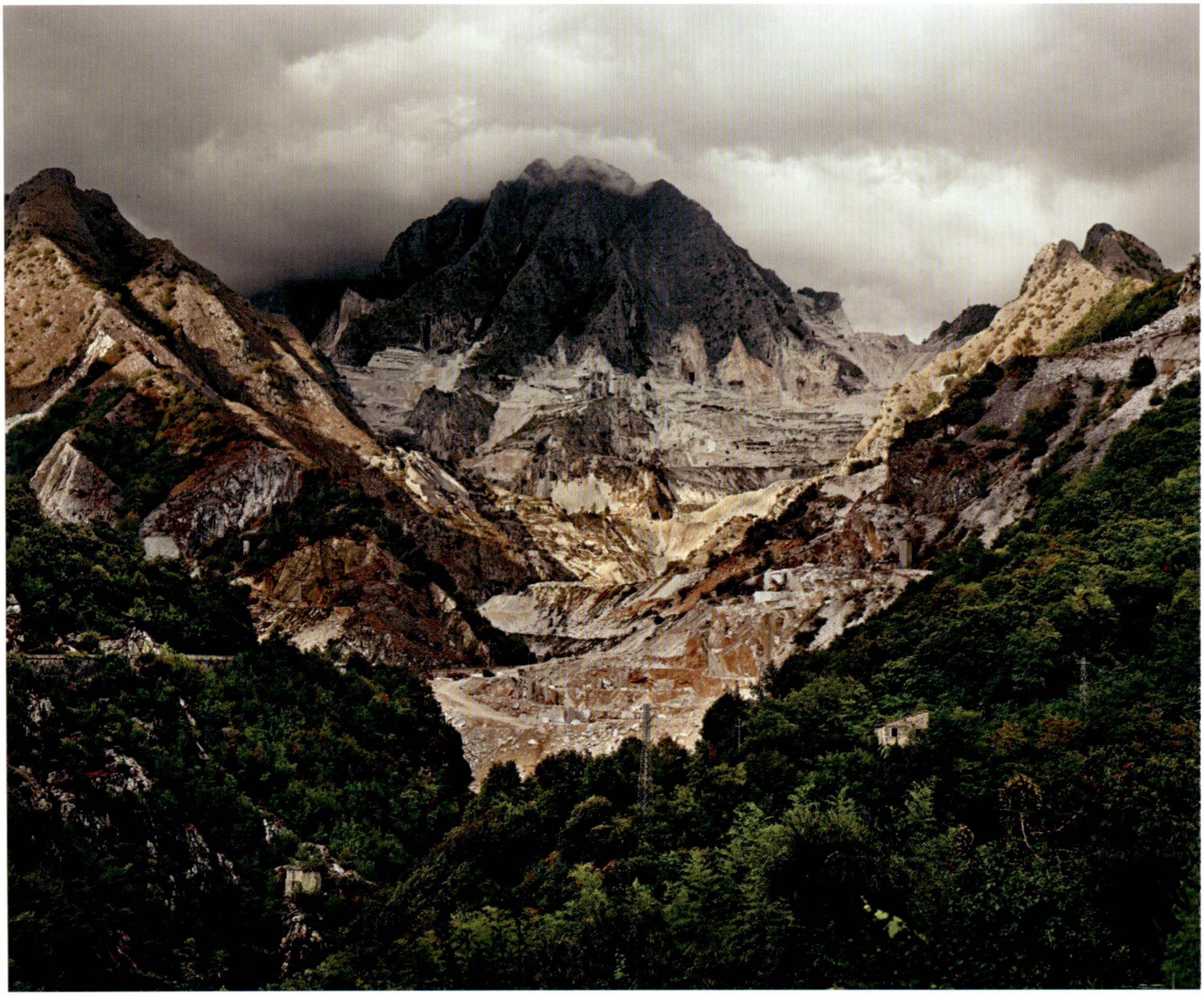

II Plates Carrara Marble Quarries

36 **Plate 7**
Edward Burtynsky
Carrara Marble Quarries #20, Carrara, Italy, 1993
Chromogenic color print
Lent by Raphael and Jane Bernstein Collection

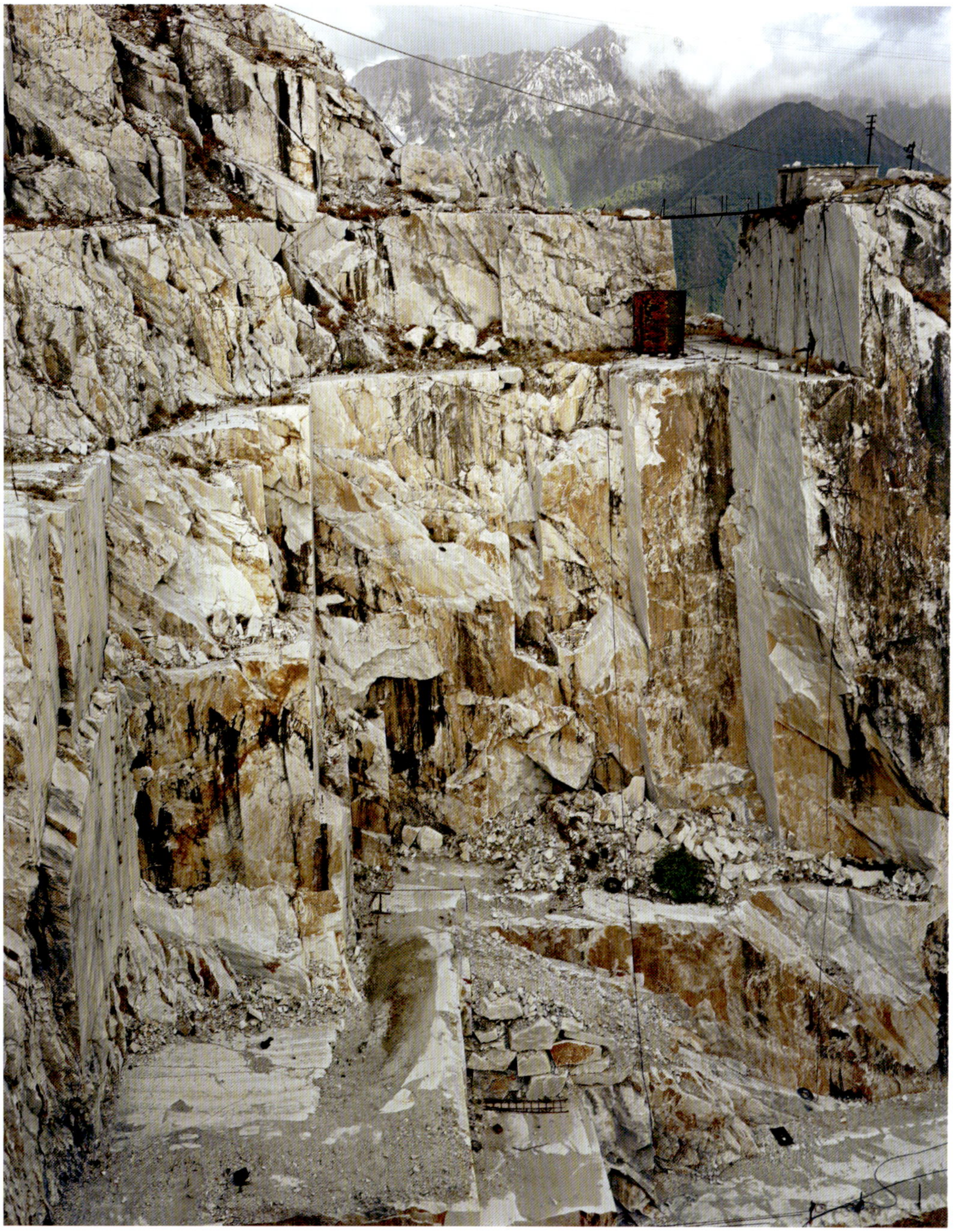

Plate 8
Edward Burtynsky
Carrara Marble Quarries #26, Carrara, Italy, 1993
Chromogenic color print
Lent by Raphael and Jane Bernstein Collection

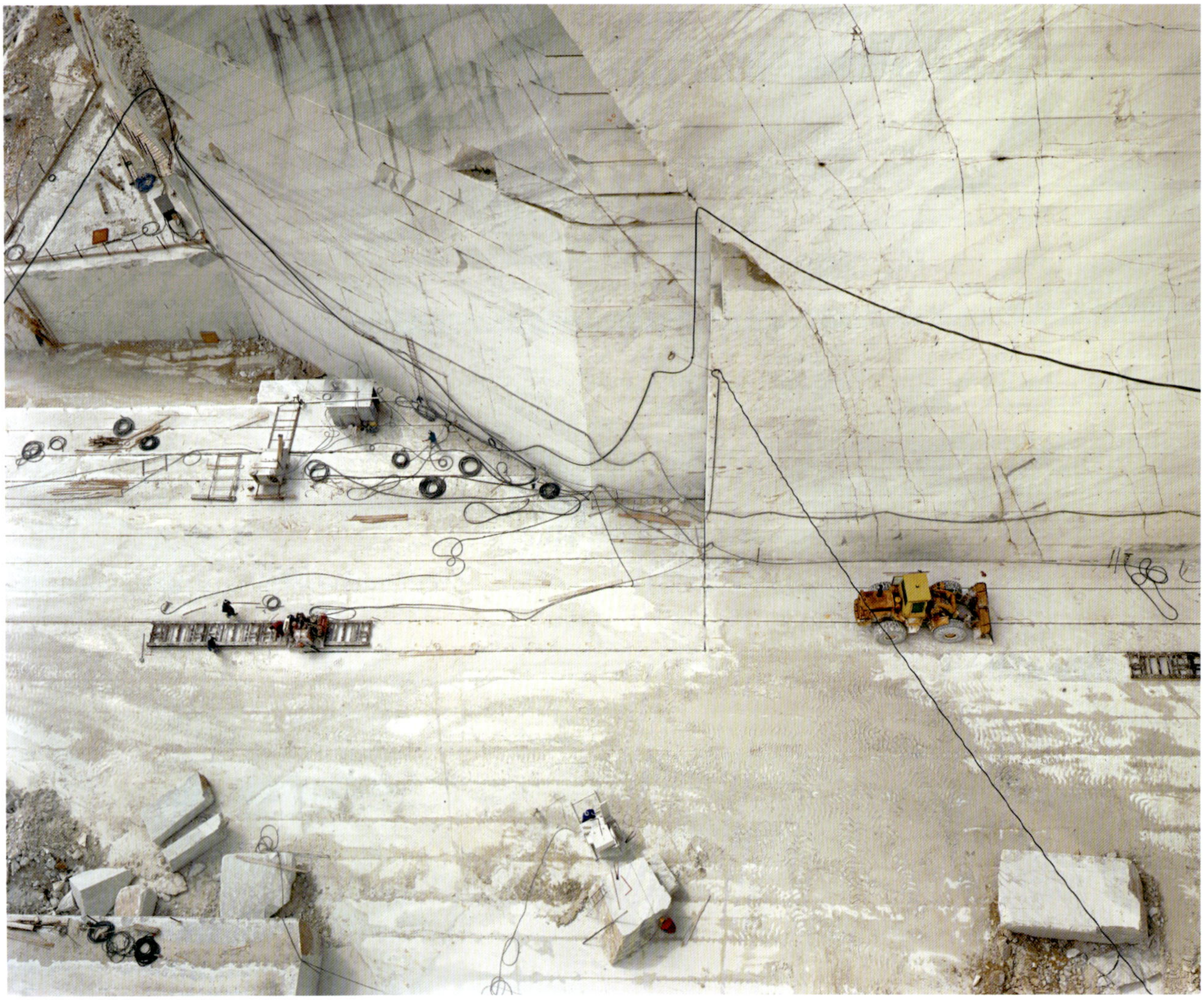

Plate 9
Edward Burtynsky
Carrara Marble Quarries #24, Carrara, Italy, 1993
Chromogenic color print
Lent by Daniel Bernstein and Claire Foerster

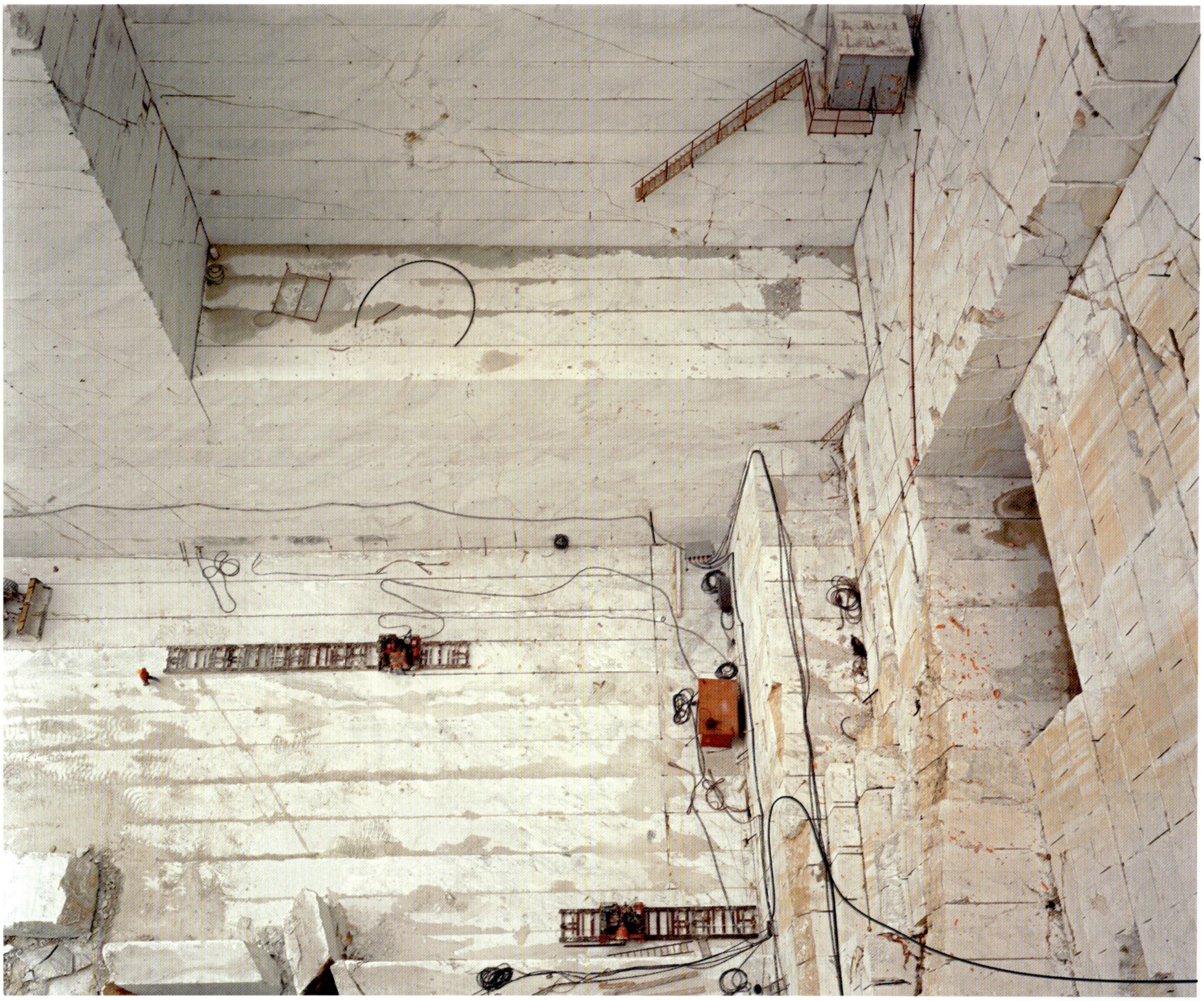

Plate 10
Edward Burtynsky
Carrara Marble Quarries #25, Carrara, Italy, 1993
Chromogenic color print
Lent by Daniel Bernstein and Claire Foerster

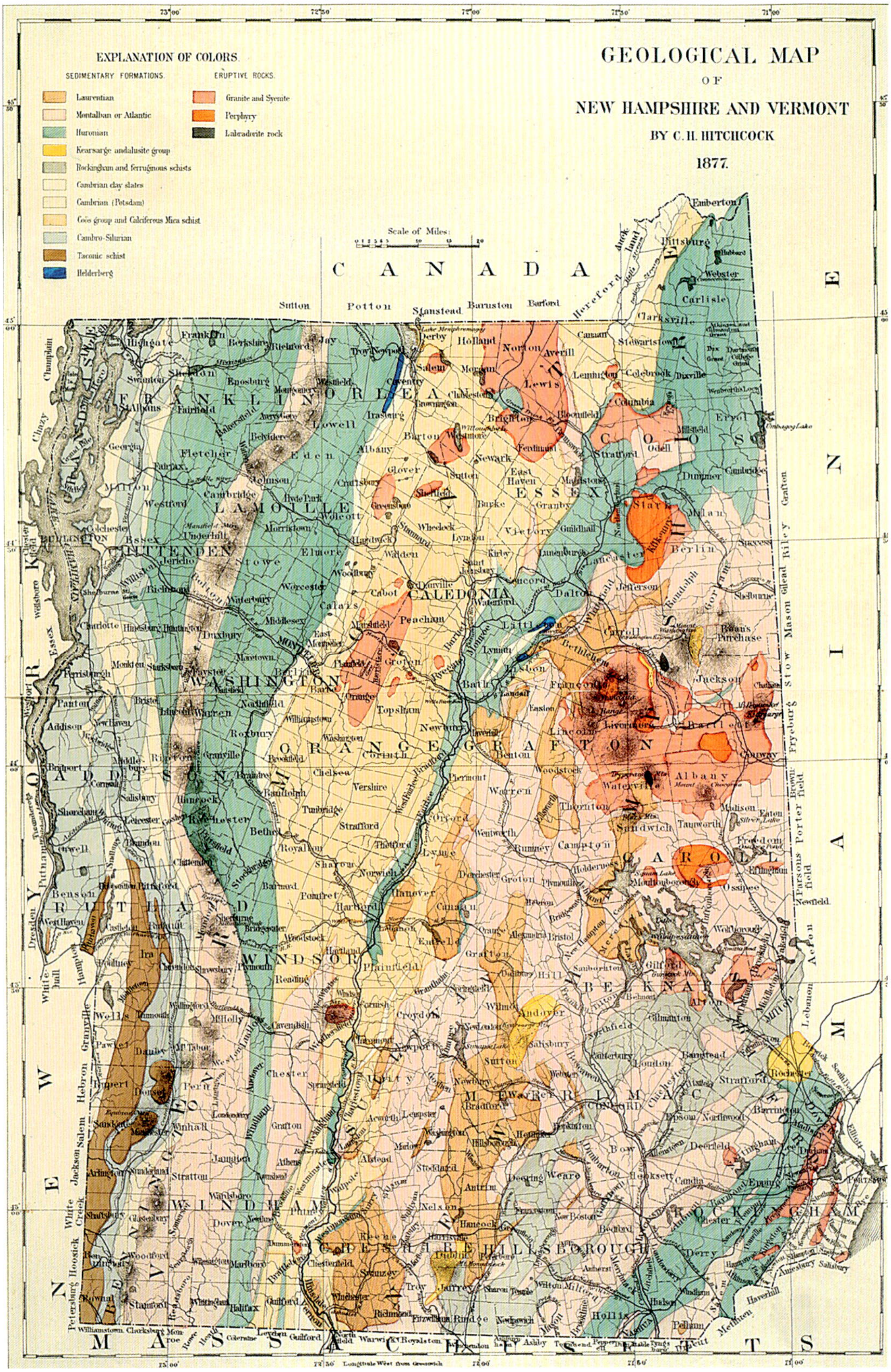
GEOLOGICAL MAP
OF
NEW HAMPSHIRE AND VERMONT
BY C. H. HITCHCOCK
1877.
EXPLANATION OF COLORS.
SEDIMENTARY FORMATIONS.
ERUPTIVE ROCKS.
Laurentian
Montalban or Atlantic
Huronian
Kearsarge andalusite group
Rockingham and ferruginous schists
Cambrian clay slates
Cambrian (Potsdam)
Coös group and Calciferous Mica schist
Cambro-Silurian
Taconic schist
Helderberg
Granite and Syenite
Porphyry
Labradorite rock
Scale of Miles.
C A N A D A
Longitude West from Greenwich

4

Appalachian Legacy

The Uncovered Record of Mountain Building in New England

Gary D. Johnson

The past three billion years of Earth's history have seen many geological and physical events that have induced significant changes in continental geography and climate. In turn, this record of continuous geological change has significantly influenced the character of agriculture, commerce, and transportation; the availability of mineral, energy, and water resources; and even the nature and composition of the communities that occupy the landscape. New England's complex of rock underpinnings, for example, has yielded an immense variety of materials that have found their way into the very fabric of society: fieldstone house foundations, the ornamental quarry stone enhancing many public and private buildings, the slated rooflines adorning New England's unique architecture, and, of course, the markers and monuments of the cemeteries, where much of this varied rock can be found. The Vermont quarry photographs of Edward Burtynsky so richly supply a visual record of this local geological history that they demand further exposition. Those areas of Vermont that have become known for their marble and granite deposits all have their origins in the episodic events of mountain building that ultimately formed the backbone of New England, and this is their story.

Opposite:
The geologic map of Vermont and New Hampshire (1877) by Charles H. Hitchcock, New Hampshire state geologist and professor of geology and mineralogy at Dartmouth College, showing the distribution of various rock sequences and their assumed ages. This map, the result of a decade of research and mapping under often difficult logistical conditions, represents a significant scientific accomplishment at a time when the distribution of natural resources was first being evaluated in the post–Civil War era. The rock underpinnings of these two states are eloquently interpreted in this prescient mapping effort.

Introduction

The landscape of New England represents a fascinating legacy of the complex changes that have taken place during the most recent billion years of Earth's history. Much of what we observe as we traverse the hills and valleys of this region is dominated by the sculpted and modified landforms that were created during the recent Pleistocene ice age and its aftermath over the last twenty thousand years or so. However, one is constantly reminded of an earlier history as well, one represented by the rock outcroppings seen in our road cuts, along our stream valleys, and on our often barren mountaintops. These underpinnings to the Pleistocene record are its raw materials, the rocks and minerals that then enable us to examine "deep time"—the events, conditions, and environmental context behind the myriad rock formations that shape New England itself (see p. 40).

The Appalachian Mountains in general, and the New England portion of this range in particular (which includes Vermont's Green Mountains and much of adjoining Vermont and New Hampshire), are the result of mountain building activity that began more than 470 million years ago. Previously existing continental landscapes and even small ocean basins were assembled, deformed, and ultimately raised into a mountain range that many believe was on the scale of the Himalayas today, the mighty result of a collision between at least two vast crustal plates. That collision started a series of events that not only produced the Appalachians but also put into place the underlying rock formations in which we quarry today.

The Geological Development of Northern New England

The geological record preserved in New England is long and complex. Within that history is an interval of some 400 million years, from roughly 650 to 250 million years ago, that saw the collision of the continents and its formidable aftermath. Some 500 million years ago, the whole of North America was still a much smaller landmass, perhaps only one-half the size it is today. Its "east coast" was actually a south-facing coastline situated at approximately 30–35°S—the present latitude of Santiago, Chile, or Cape Town, South Africa. Many changes in this continental configuration have taken place since, including the subsequent assembly and disassembly of several continental masses and their eventual migration to the positions in which we find them today.

The position of our continent by 470 million years ago, and its relationship to the surrounding seas and other continents, has been interpreted as the final stage of an ocean basin expansion that likely began roughly 200 million years before, at the end of the time period referred to as the Precambrian. At that distant time, North America had just broken away from a mix of continental masses that had been assembled beginning about one billion years ago and had included parts of northwestern Africa, South America, Australia, and Eurasia. In each case, these separating landmasses were geographically much different then than they are today.

Off the "eastern" seaboard of North America at that time, we would have seen an ocean basin that continued to widen until tectonic forces began to change the geometry of its boundaries with our continent. This ocean, known as Iapetus, played a pivotal role in the early geological development of both North America and Europe. Roughly 470 million years ago, Iapetus began to close, and tectonic forces began to bring the opposing shorelines of North America (known as Laurentia after the Laurentian highlands of eastern Canada) and northwestern Europe (known as Baltica after the landscape surrounding the present-day Baltic Sea) progressively closer together in a complex scenario that took nearly 100 million years to complete.

Oceanic closure, which is generally characterized by a process that we call subduction, results when one tectonic plate is thrust beneath an adjacent tectonic plate. In this instance, the lower (or subducting) plate was comprised of the oceanic crust of the Iapetus, a fifteen- to twenty-kilometer-thick sequence of relatively dense rocks that included basalts. These had been formed by erupting magma within oceanic settings, a process still going on today in Iceland, which straddles the Atlantic mid-ocean ridge, or so-called spreading axis, from which the modern continents of North America and Europe are presently parting. The upper plate in this subduction scenario was comprised of a continental crust that was some forty-five kilometers thick and made up of relatively light, lower-density rocks that included granites and metamorphosed sediments. The interface between these two plates, or the subduction zone, eventually resulted in the establishment of an uplifted volcanic arc that was somewhat akin to the islands of Japan or the Philippines in today's world.

Perhaps one of the most unusual events associated with the subduction process was the development of the Taconic Range of western Vermont and Massachusetts and eastern New York. By some accounts, the rock succession that makes up this range was derived originally from a series of tectonic "slices" from a position farther to the east. It is now more commonly believed, however, that these displaced rocks may have originated from a position along what is now the Connecticut Valley region of New England, and that they were displaced as a result of westward thrusting from the rising volcanic arc to the east during these initial stages of the Iapetus closure. The emplacement of this massive series of tectonic slices subsequently had a profound effect on the sediments that they came to overlie. Previously deposited shallow marine limestone, some of which records the earliest development of reef-building organisms, became metamorphosed during this burial episode into the marbles of western Vermont that we see exposed today (fig. 4.1).

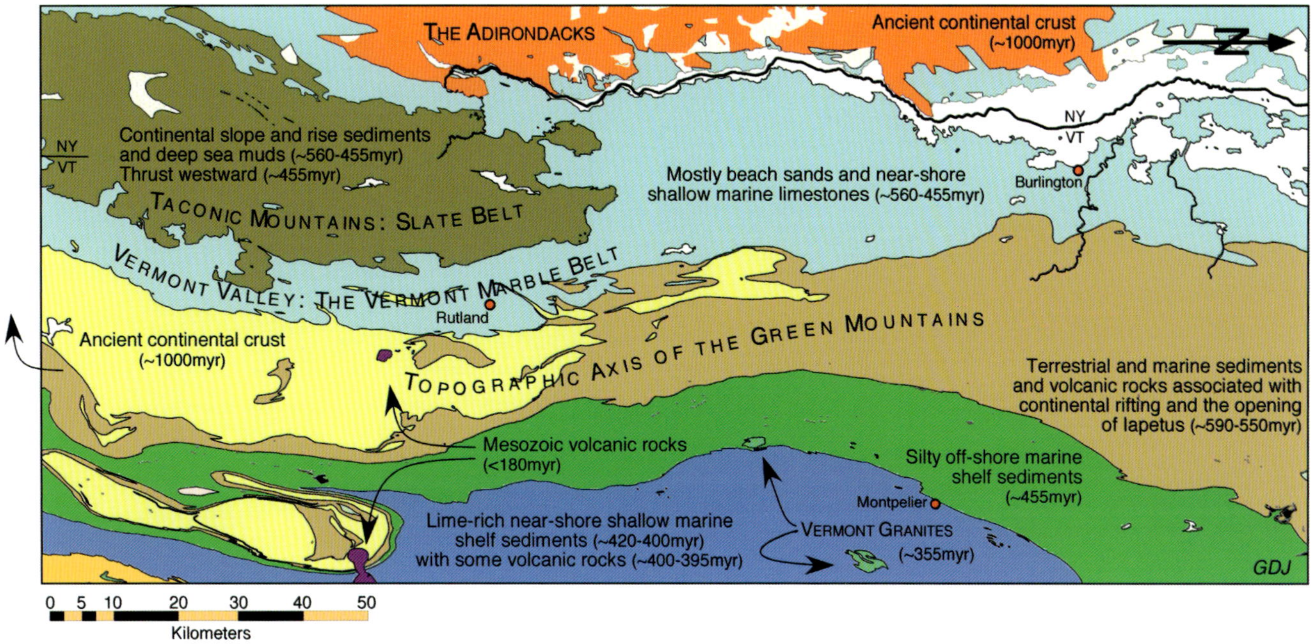

Figure 4.1

The Legacy of Iapatus: Mountain Building in Vermont. A geologic sketch map of the western portion of Vermont illustrating the distribution and various rock sequences that make up the Vermont portion of the Appalachian Mountains. The location and areal extent of the Vermont marble belt, the Vermont slate belt, and various granite occurrences are also illustrated. Modified and reinterpreted from B. Doolan, *Rocks and Minerals*, vol. 71 of *The Geology of Vermont* (1996), pp. 218–25.

The structural displacements accompanying the initiation of closure of the Iapetus Ocean now represent the tectonic phase known as the Taconic "Orogeny," a nearly 200-million-year assembly of the formerly disparate elements of that late-Precambrian continental mass. During this phase, Iapetus closed. An offshore archipelago referred to as Avalonia after the Avalon Peninsula of Newfoundland (where its uniqueness was first recognized) came into contact with North America, with northwestern Europe, in the form of the so-called Baltica terrane, soon to follow. This sequence of events led up to what is known as the Acadian Orogeny (beginning roughly 370 million years ago) after the northernmost portion of New England and maritime Canada, or Acadie (fig. 4.2).

This Acadian Orogeny is the nearly final and most important geologic contribution to the dramatic development of the New England Appalachians. By this time, our mountains probably exceeded seven kilometers in elevation (remember those Himalayas?). Such tall mountains also have very deep roots, which create metamorphic conditions through extremely high temperatures and high pressures at depth approaching 50 kilometers or more within the earth's crust. This metamorphic environment induces regional changes in mineral composition and the very fabric of the rock; coupled with materials associated with fluids passing amongst these rocks, entirely new rock materials can be formed as well. Limestones become marbles, shales become slates or even schists, and, at the extreme, some of these rocks—which were originally the sediments of Iapetus itself—melt, to likely become, upon their eventual cooling, the granites for which parts of New England have become famous.

Since these mountain-building events were completed, some 250 million years of erosion, coupled with a continuous (and compensating) uplift of the New England crust, have taken place. The sediments derived from these eroding mountains have contributed to the growth of the continental shelves that border the modern Atlantic Ocean. The rock assemblages that formerly were at extreme depth as the roots of the Appalachians are now outcrops on the surface and signature characteristics of the landscape that we have

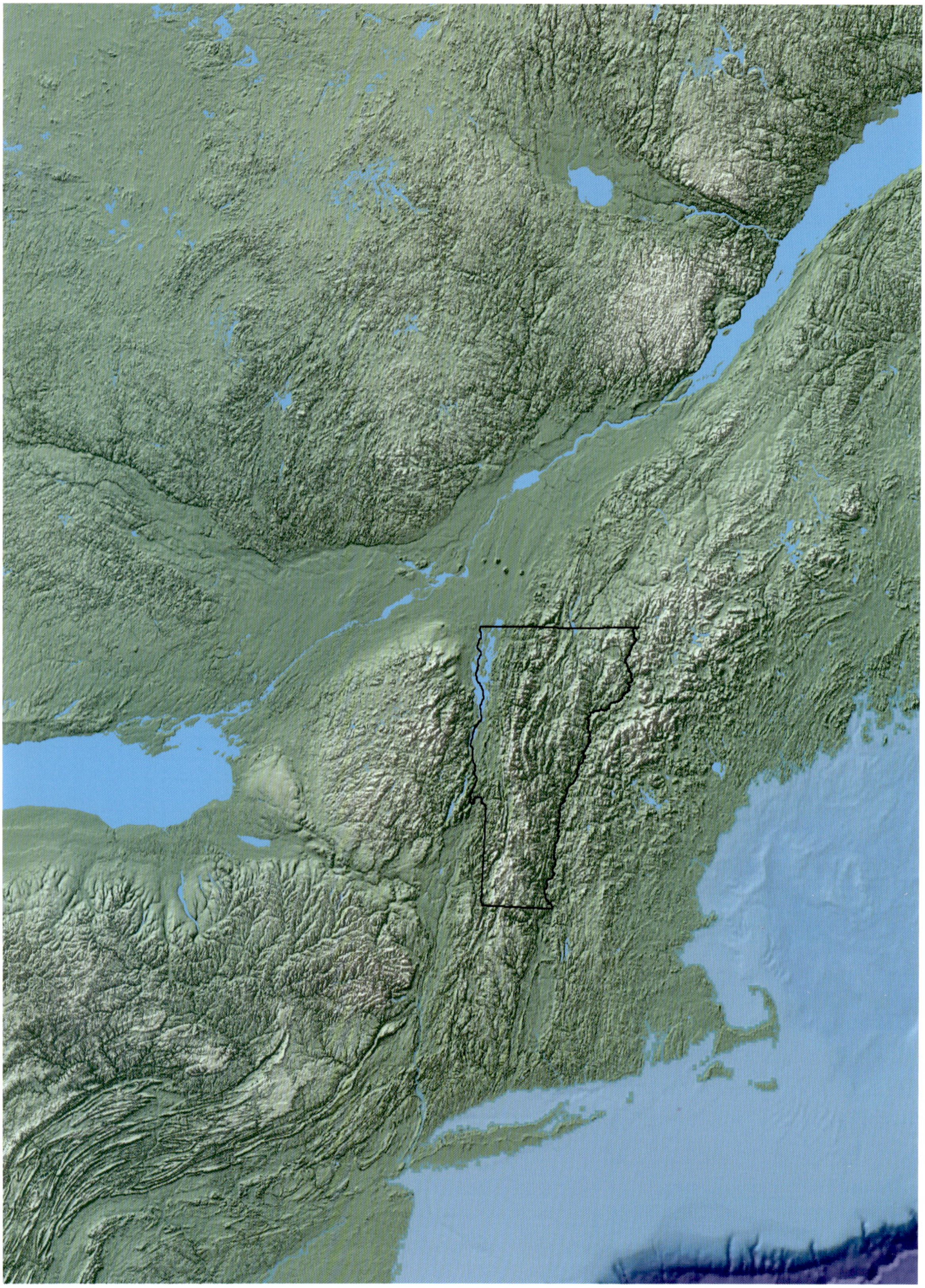

Figure 4.2
A shaded relief representation of the topography of New England and the maritime provinces of Canada. This illustrates well the roughly north-south linear fabric of the Appalachian Mountains of Vermont and New Hampshire as well as their relatively narrow east-west continuity, which was inherited from the compressional forces that shaped the Appalachian uplift. Graphic by Jonathan Chipman.

come to consider quintessentially New England—that is, the raw materials, including the marbles and granites of Vermont, that we quarry today.

The Fabric of Rock: Architecture, Textures, and Geometry

The utilization of natural materials as architectural or ornamental components extends back into our earliest prehistory. Much more recently, of course, locally quarried materials have been used to form the foundations, façades, and roofing elements of many of the buildings that have filled our modern landscape. In many respects, the use of these materials in New England towns offers significant insights into the quarrying operations, transportation systems, and material availability of a given era. For example, the nucleus of many villages is comprised of buildings that utilized exclusively local materials in their construction. As these communities grew, however, newer buildings that developed on the periphery of this nucleus often used materials that were both local and imported from more distant locales, as transportation capabilities improved.

A case in point is the Dartmouth College campus and its surrounding community. The oldest buildings, from the late eighteenth and early nineteenth centuries, have deep foundation stones (fig. 4.3) that are either glacial erratics collected on the surface from the surrounding hills or rough quarry stones collected only hundreds of meters from the eventual site of these buildings.[1] To complete the aboveground portions of these footings, the topmost foundation elements were usually hand-finished and hand-dressed stones from nearby quarries. Transportation to and from these local sites was by draft animals hauling wagons or rock sleds, an obviously impractical means of securing building materials from more distant sources.[2]

From the early nineteenth century until the post–Civil War era, most of the natural materials used in construction remained local. With the advent of the privately developed turnpike road system, the development of publicly funded thoroughfares, and the expansion of rail lines into many of the valleys of Vermont and the adjoining areas, an opportunity arose to augment locally sourced building materials with, among other things, the products of the rapidly developing quarrying operations of western Vermont. No longer would locally quarried, hand-finished stone limit the character of a building's thresholds, lintels, sills, and so on, and newer Dartmouth buildings such as Webster Hall and the Butterfield Museum (replaced by Baker Library in 1928) saw the increased utilization of materials brought west from quarrying operations in Grafton or even Concord, New Hampshire. Improved access to the marble, granite, and slate quarrying operations in central and western Vermont brought an increased use of these materials as well, with slate for roofing purposes proving to be particularly popular early on.[3]

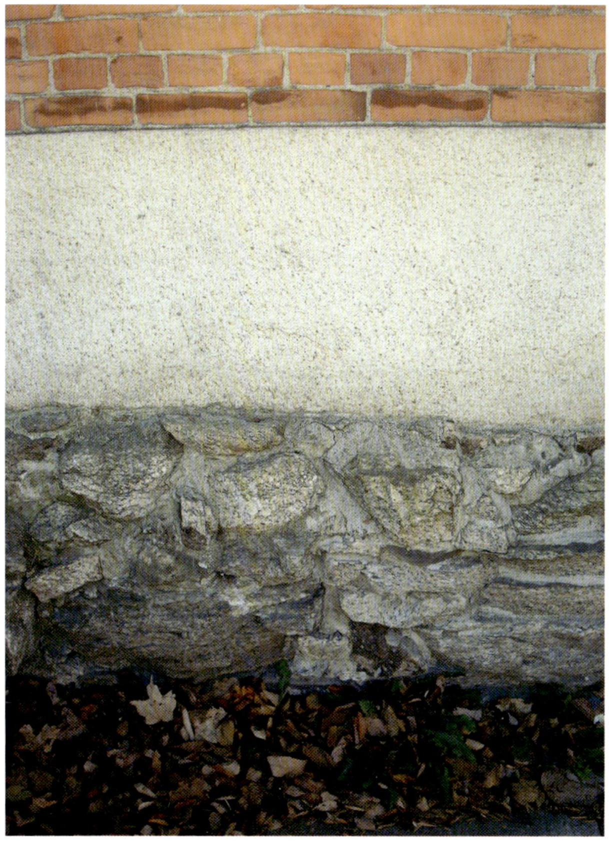

Figure 4.3
A typical early foundation construction style in which a basal course of field stone—often locally claimed glacial rubble—is capped by quarried granite beams that spread the load of the overlying brick wall. In this example, a wall is subsequently laid directly on the quarried granite beam.

Rock quarrying operations throughout New England, and New Hampshire and Vermont in particular, benefit from the wide array of rock types, rock fabrics, and geological structures that represent the legacy of Appalachian mountain building those hundreds of millions of years ago. And all of these signature stones—the limestone, the marble, the slate, the Verde antique, the talc, and the granites that occupy the quarrying operations of our states—share that link to the ancient Iapetus. The components of a rock's fabric—the size, shape, arrangement, and chemical composition of its individual mineral grains—are amongst the most important properties that constrain the utilization of these materials.[4] While it is less important to the stone used as large rubble to stabilize embankments against erosion or as wall material, this fabric has a much greater effect upon other applications.[5]

The various marble quarries located west of the Green Mountains in Vermont represent a highlight of the natural resources available to the local and regional market. Like the famous marble quarries of Carrara, Italy, that for centuries have supplied both builders and artisans with an easily worked material, many of the marble deposits of Vermont are of extremely fine quality. With marble, which is the metamorphic result of subjecting limestone to high temperature and pressures, the original fabric of the limestone is unrecognizably altered. While the chemical composition remains primarily the same, the size, shape, and arrangement of the original calcite (calcium carbonate) or dolomite (calcium magnesium carbonate) crystals are altered. Ideally, this metamorphism will result in a rock of uniform texture, with little or no variations in its internal structure.

The metamorphosed sediments pictured in *Abandoned Marble Quarry #18* by Edward Burtynsky (plate 2) illustrate the complex fabric of the marble, which may be further enhanced by decades of exposure to the elements, which highlights irregularities through differing rates of weathering. Certain structural properties such as fractures and joints can help to enhance this response as well. The ultimate result is an easily worked material, relatively soft and readily manipulated with iron cutting tools but revealing no discernable zones of weakness. It is ideal for sculpture and historically used for tombstones (primarily before the Civil War), monuments, and ornamental architectural elements in building construction. The only serious drawback is that this material is highly reactive while weathering and as such can deteriorate significantly over time. A brief assessment of grave markers in virtually any cemetery in New England will quickly demonstrate that most of the marble burial markers are becoming illegible—and this after only a century or so (fig. 4.4). Like the sediments that make up the slate belt shales, the metamorphosed limestone of the Vermont "marble belt" was likewise deposited along the margins of the Iapetus, although much closer to its very shallow marginal portion. These marble-belt sediments are exposed today because the formerly

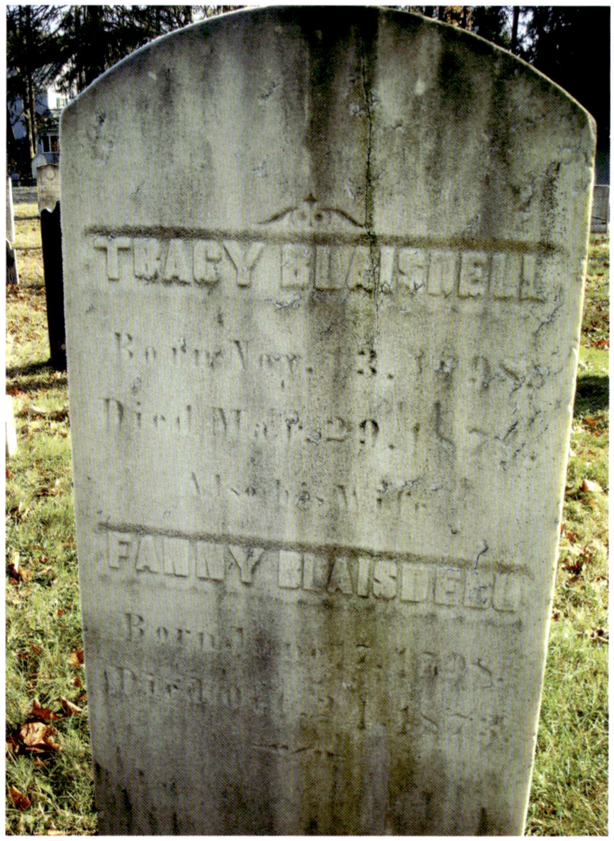

Figure 4.4
A severely weathered marble headstone from the late nineteenth century (about 1873) exhibiting significant loss of detail as compared to contemporary granite headstones (see fig. 4.6). Fairview Cemetery, Norwich, Vermont.

overlaying Taconic sequence of displaced strata has been eroded away.

The most important quarrying operations in northern New England—and New Hampshire and Vermont are no exception—relate to the extraction of granite. Not all of these granites are actually granitic in composition (that is, containing at least 20 percent quartz and a large percentage of the alkaline, potassium-rich feldspar known as orthoclase, which gives the rock a pinkish cast). Instead, many of these rocks are actually granodiorites, in which a different type of feldspar, a sodium and calcium-rich form known as plagioclase, is dominant (fig. 4.5). These granodiorites are more commonly grayish in color. It is no mystery that Vermont is recognized as being host to one of the most desirable granites in the United States: in terms of the rock's fabric, the desirable properties of granite (or granodiorite) must again be its durability, workability, and general lack of compositional or structural imperfections, and Vermont's deposits reveal the very same—stone that is of relatively uniform grain size and mineral distribution and that lacks any mineralogic concentrations that could compromise its structural integrity.

Despite its quality, however, challenges continue to accompany its extraction. The quarry face in *Rock of Ages #4* by Edward Burtynsky (plate 12) illustrates the problems associated with fractures that may exist within the rock mass that is being quarried. In this example, most of the many near-horizontal and the several near-vertical fractures came about as a result of the natural removal of many kilometers of overlying rock over millions of years. This "unloading" enables the rock mass to expand ever so slightly, resulting in the fracture system that is so well illustrated here. Additionally, the load exerted by up to a mile of ice accumulation during the recent Pleistocene ice age, followed by its relatively rapid melting and loss over only several thousand years, caused these rocks to fracture even further. Although portions of this quarry certainly can provide high-quality stone for a variety of applications, only those smaller masses bounded by these large fractures can be utilized.

Although Vermont has abundant granitic or

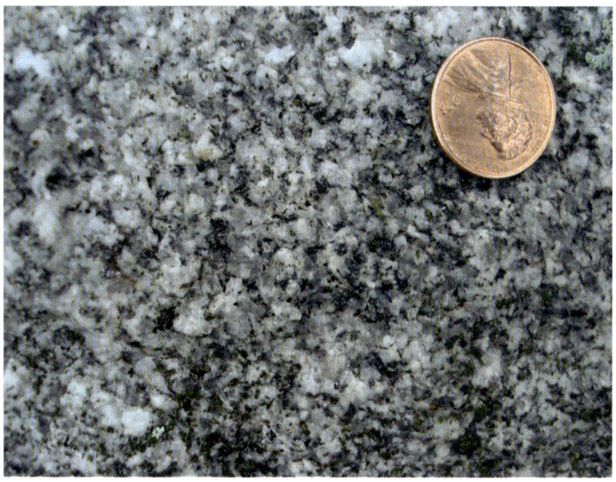

Figure 4.5
Barre "granite" in outcrop. The uniformly mixed mineral assemblage of this rock is readily evident in this exposed rock face. The dark minerals of biotite and hornblende, the milky, somewhat translucent quartz, and the white plagioclase feldspar typically characterize this well-known quarry rock. From the Rock of Ages Quarry, Upper Graniteville, Vermont.

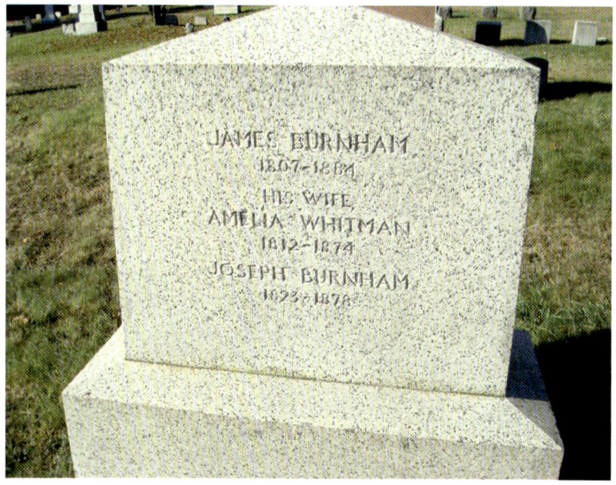

Figure 4.6
A typical granite headstone detail from the late nineteenth century (about 1884) still exhibits relatively good detail in the hand-carved inscriptions. The fresh appearance of this quarried stone is related to the enduring character of granite, as opposed to the easily weathered marble (see figure 4.4). Hillside Cemetery, Norwich, Vermont.

granodioritic rock types, not all are suitable for ornamental or monumental use.[6] The so-called Barre Granite is actually a fine-grained granodiorite, and portions of this Barre granodiorite body are of such high quality, and so free of blemish, that they represent one of the best examples in the world (fig. 4.6). The high quality of materials such as these has enabled a regional (and extra-regional) industry to flourish for nearly two centuries in this part of New England. Its origins, deep in the throes of the Earth's formative geological upheavals, are as compelling as the stone presently sustaining it.

Notes

1. At Dartmouth College for example, buildings including Dartmouth Hall, the Shattuck Observatory, and the Casque and Gauntlet Society, among others, exhibit this traditional style of foundation construction.
2. In addition to quarry stone, bricks were a popular building material in much of New England, and they presented somewhat lesser challenges as well. Proximity to the virtually unlimited fuel resource represented by nearby forests meant that local kilns could be established almost anywhere there was a need for bricks. This was, then, a locally derived resource, with very little requirement for material to be imported from any great distance.
3. By the end of the Civil War, materials that originated in the southern Connecticut River Valley had begun to appear throughout New England. The post-1850's "brown stone" architectural movement that crept into the industrial heart of many of the larger East Coast cities eventually found its way north along the Connecticut and Merrimac Rivers to the communities of northern New England as well. Red sandstones were quarried from massive operations in Massachusetts and Connecticut and then transported by rail, either as finished architectural elements or as unfinished blocks, throughout the region.
4. Improved transportation possibilities, of course, first allowed for the specification of stone based upon its usefulness rather than simply its nearness. This, in turn, improved not only structural durability but also influenced architectural taste.
5. Roofing slates, for example, to this day manufactured by hand in the area around Fair Haven, Vermont, can only be selectively quarried from deformed rock sequences in which the original layered sedimentary rock (a fine-grained shale that has metamorphosed to a compact slate) has been tightly folded such that the original sediment layers are now deformed parallel to the plane of folding. This rock then cleaves along those sediment layers without any interference from the geometry of the fold. The ease of cleaving, and the thinness of the sheets of slate that can be produced, results in a material that can successfully be used as tiles and roofing slates. Interestingly, the Vermont "slate belt" occurs within those metamorphosed shales of the Taconic sequence that had been thrust or slid westward during those early phases of closure of the Iapetus Ocean.
6. Curbing stones are currently the principal product of the Concord granite quarries in New Hampshire, but central Vermont's fine-grained granodiorite lends itself to numerous other applications.

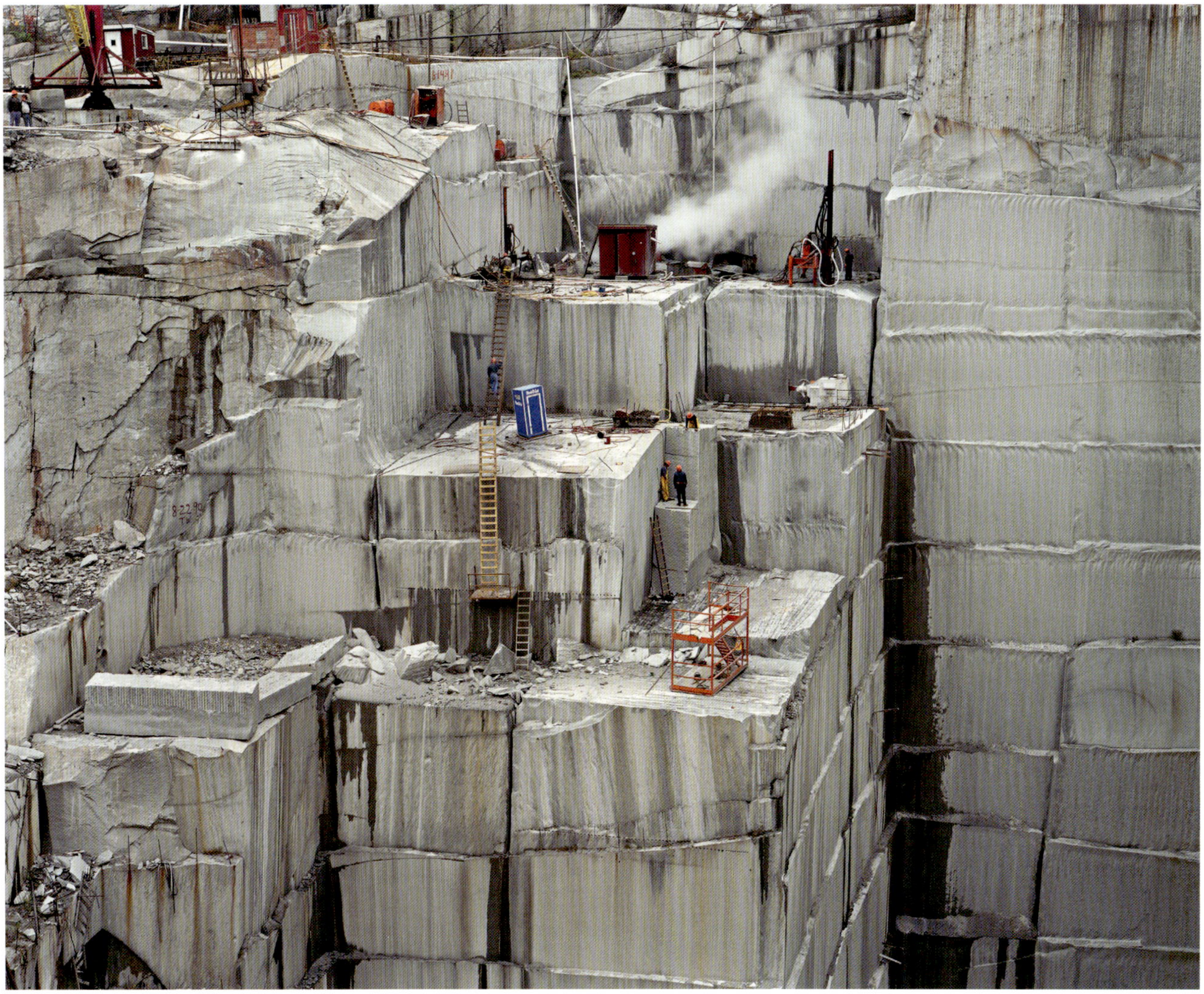

III Plates Vermont Granite Quarries

Plate 11
Edward Burtynsky
Rock of Ages #1, Active Section, E. L. Smith Quarry, Barre, Vermont, 1991
Digital chromogenic color print
Courtesy of the artist

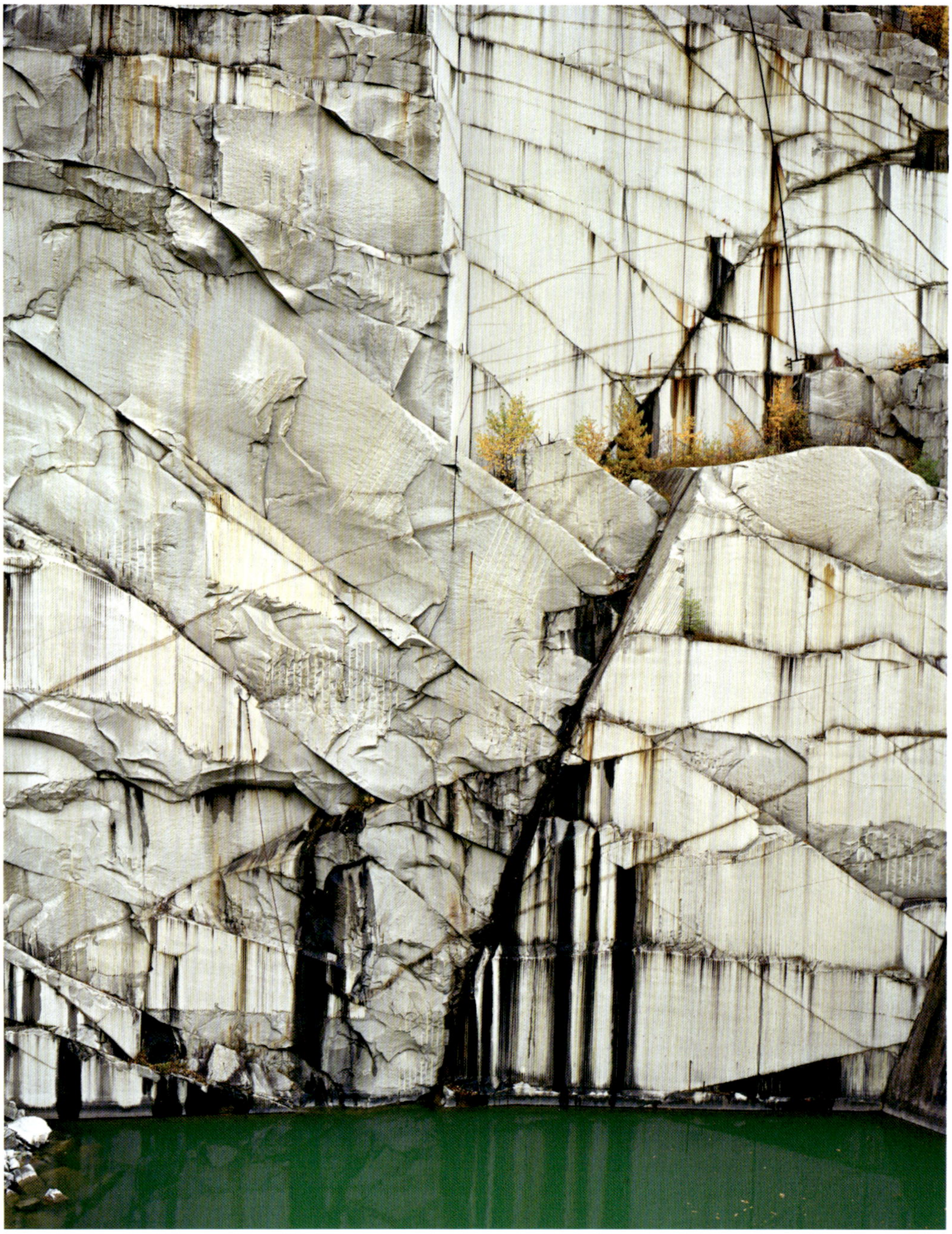

Plate 12
Edward Burtynsky
Rock of Ages #4, Abandoned Section, Adam-Pirie Quarry, Barre, Vermont, 1991
Digital chromogenic color print
Courtesy of the artist

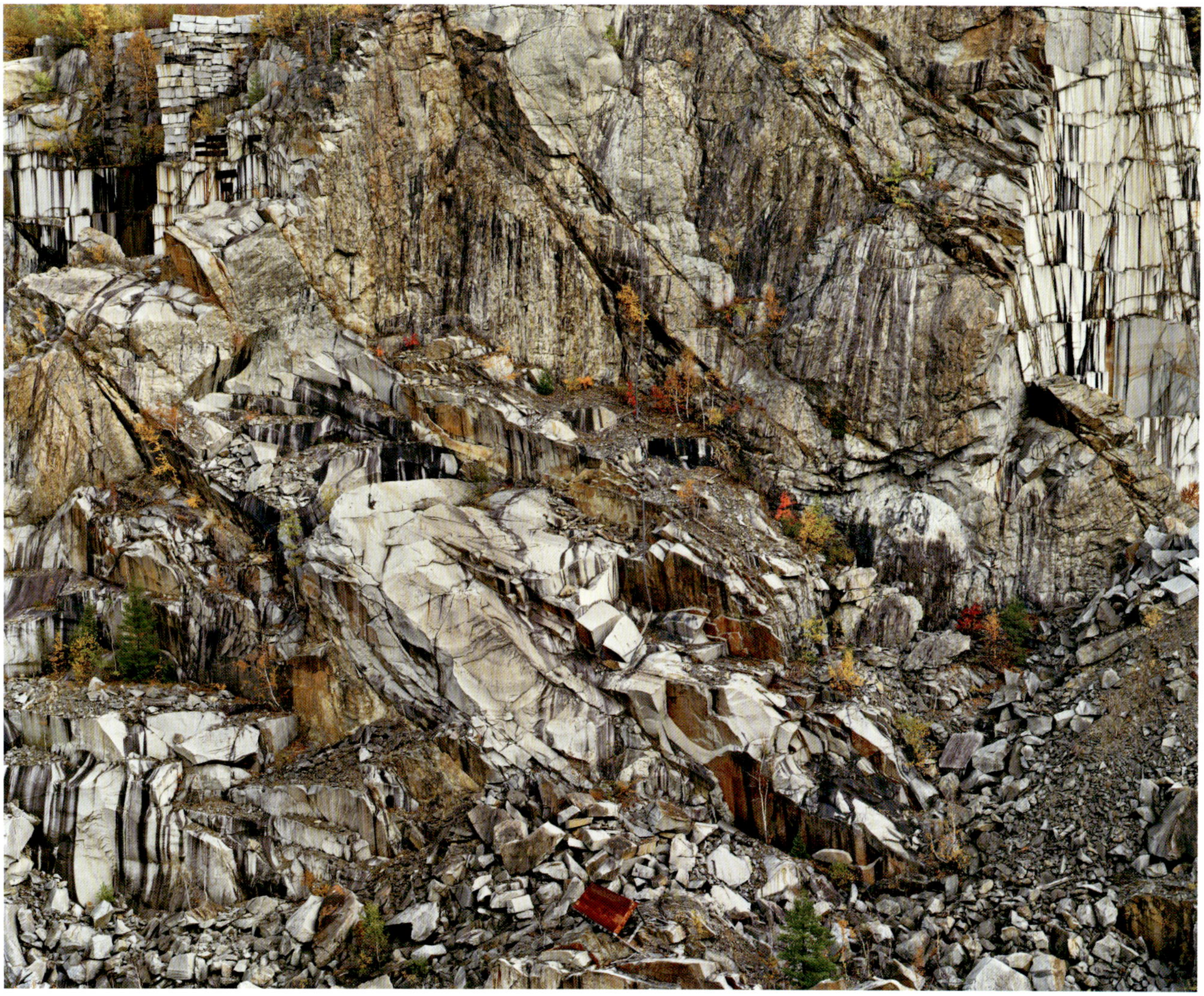

Plate 13
Edward Burtynsky
Rock of Ages #6, Abandoned Section, Rock of Ages Quarry, Barre, Vermont, 1991
Digital chromogenic color print
Courtesy of the artist

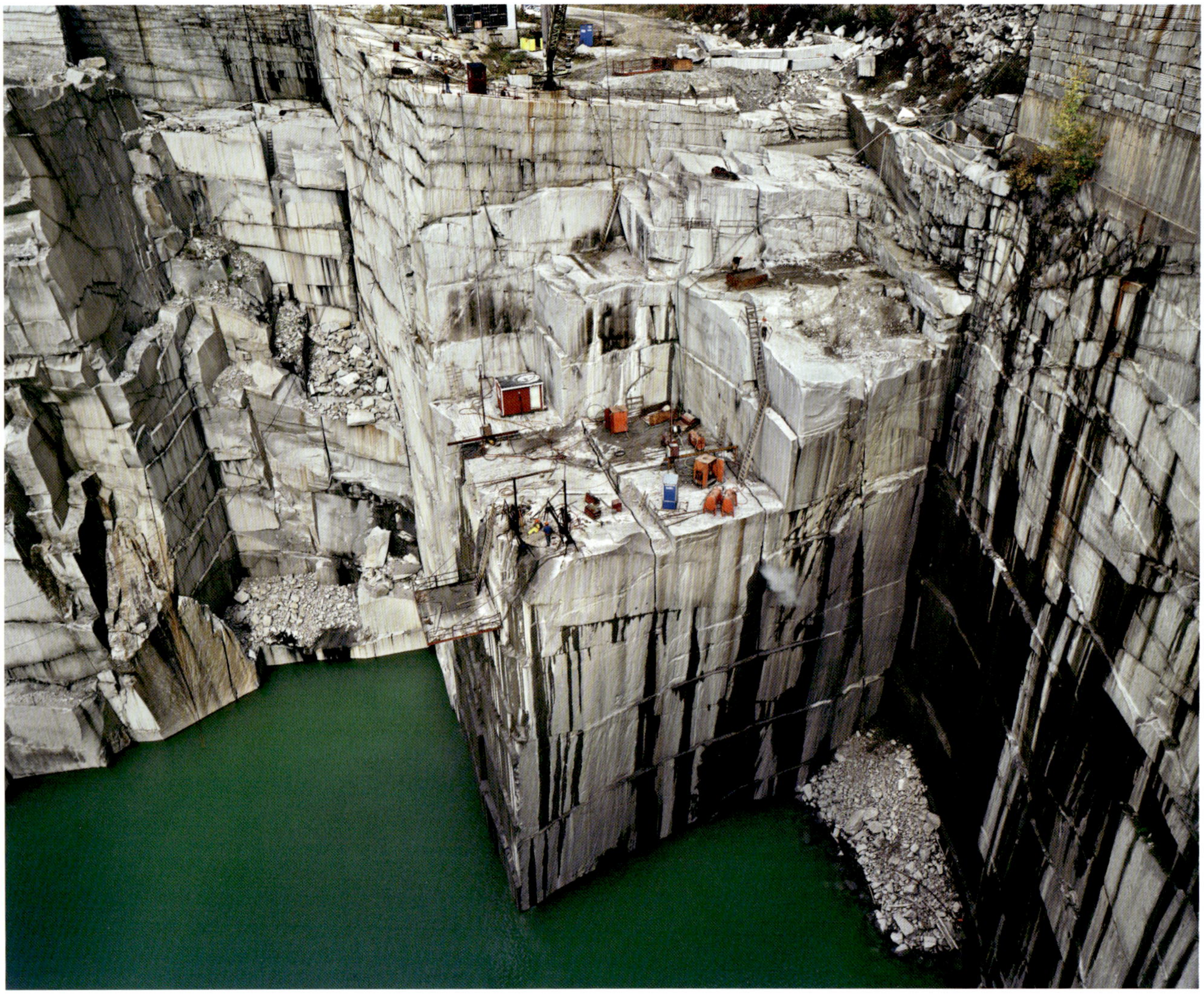

Plate 14
Edward Burtynsky
Rock of Ages #7, Active Granite Section, E. L. Smith Quarry, Barre, Vermont, 1991
Digital chromogenic color print
Courtesy of the artist

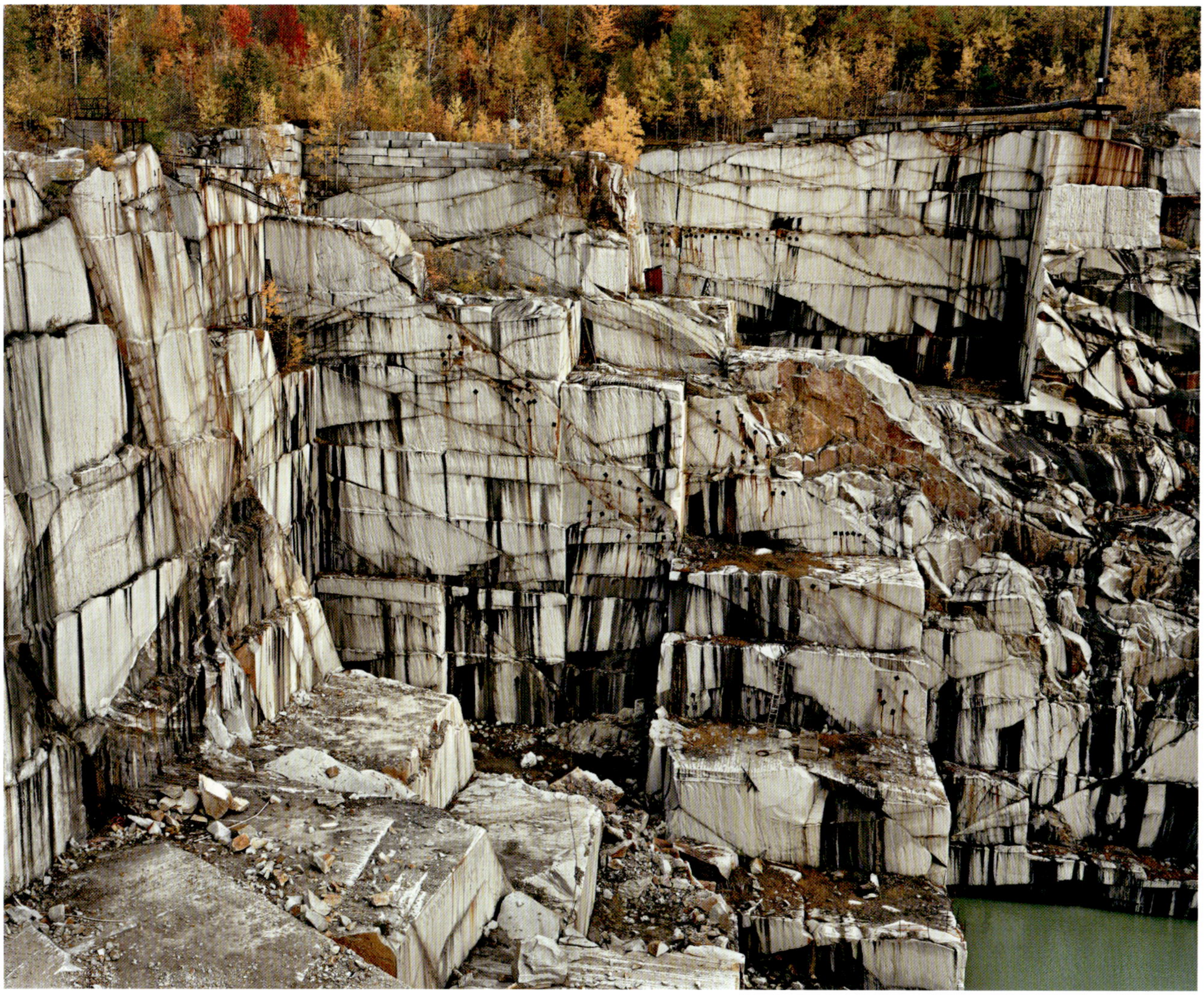

Plate 15
Edward Burtynsky
Rock of Ages #12, Abandoned Section, Adam-Pirie Quarry, Barre, Vermont, 1991
Digital chromogenic color print
Courtesy of the artist

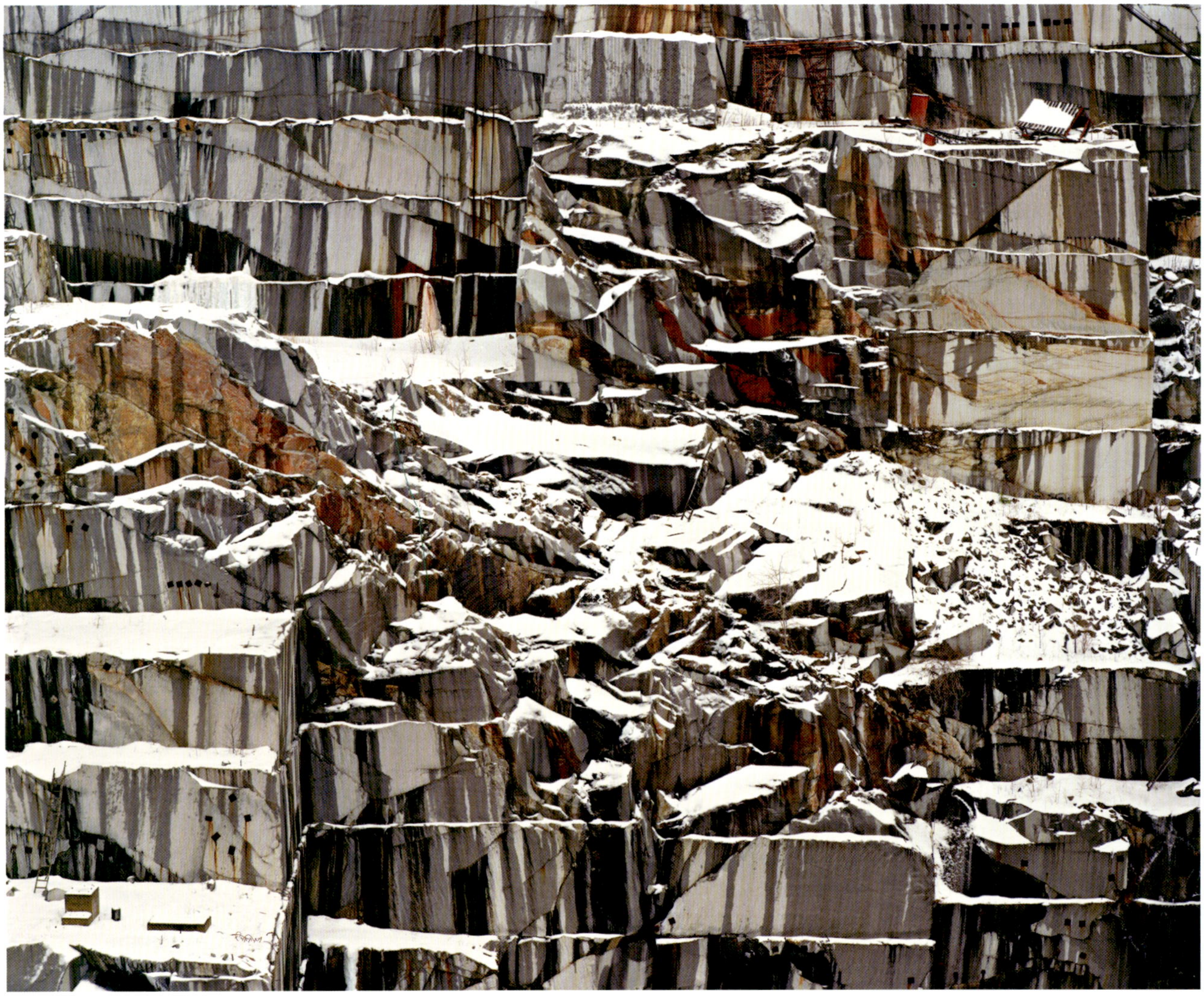

Plate 16
Edward Burtynsky
Rock of Ages #17, Abandoned Section, Adam-Pirie Quarry, Barre, Vermont, 1991
Digital chromogenic color print
Courtesy of the artist

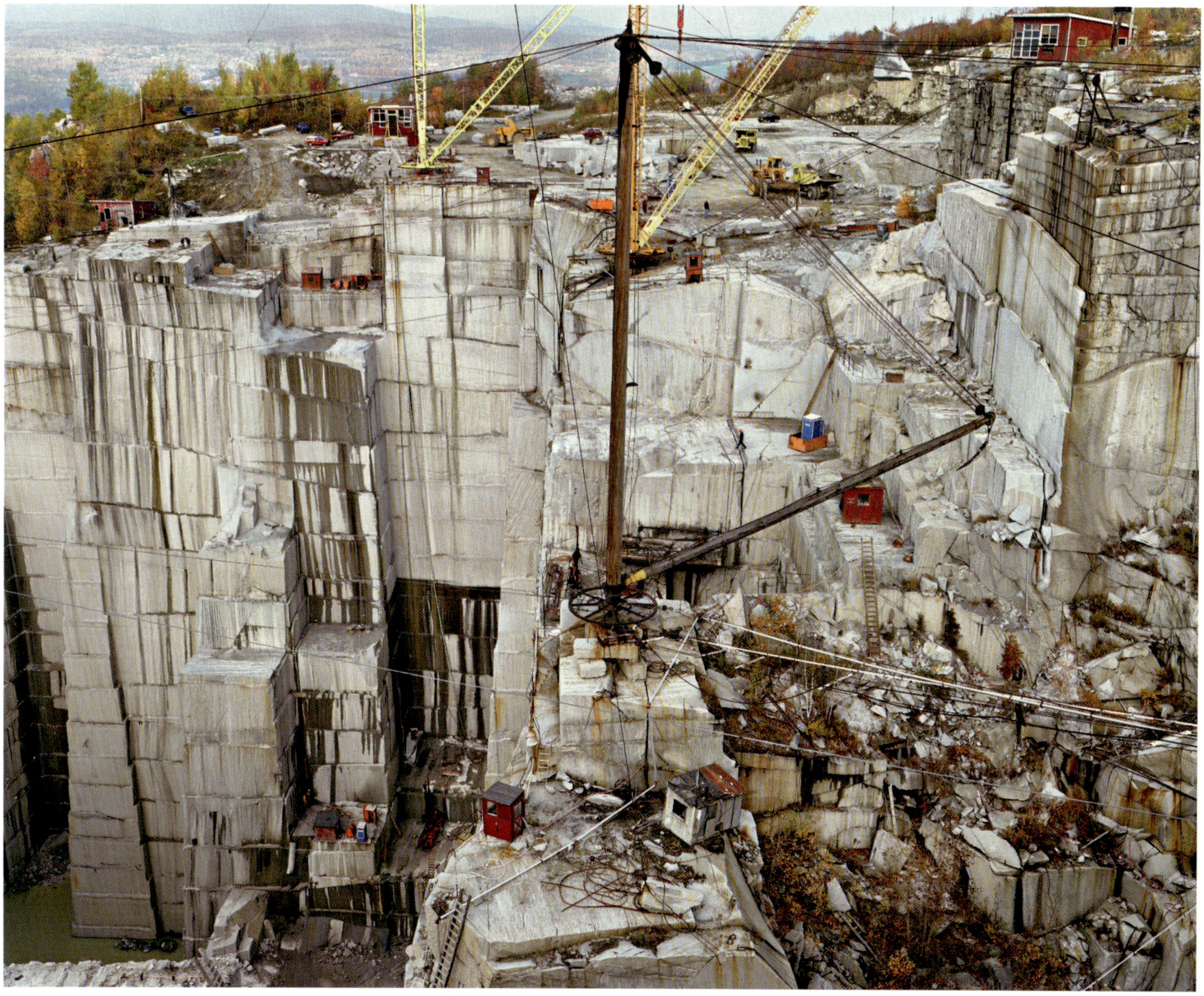

Plate 17
Edward Burtynsky
Rock of Ages #19, Granite Section, Rock of Ages Quarry, Vermont, 1991
Chromogenic color print
Lent by the Middlebury College Museum of Art, Vermont: Purchase with funds provided by the Walter Cerf Art Fund; 2008.034

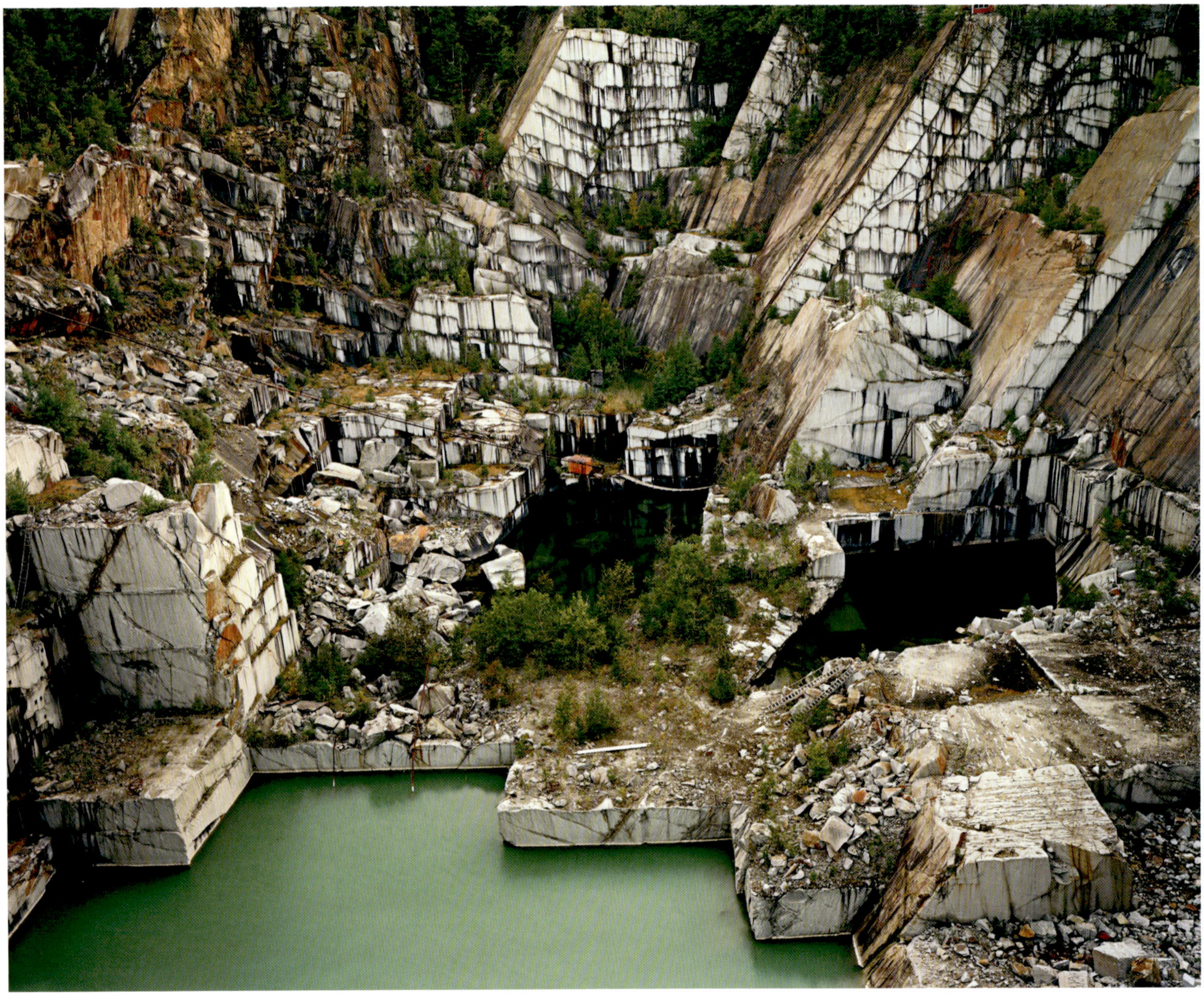

Plate 18
Edward Burtynsky
Rock of Ages #25, Abandoned Section, Adam-Pirie Quarry, Barre, Vermont, 1991
Digital chromogenic color print
Courtesy of the artist

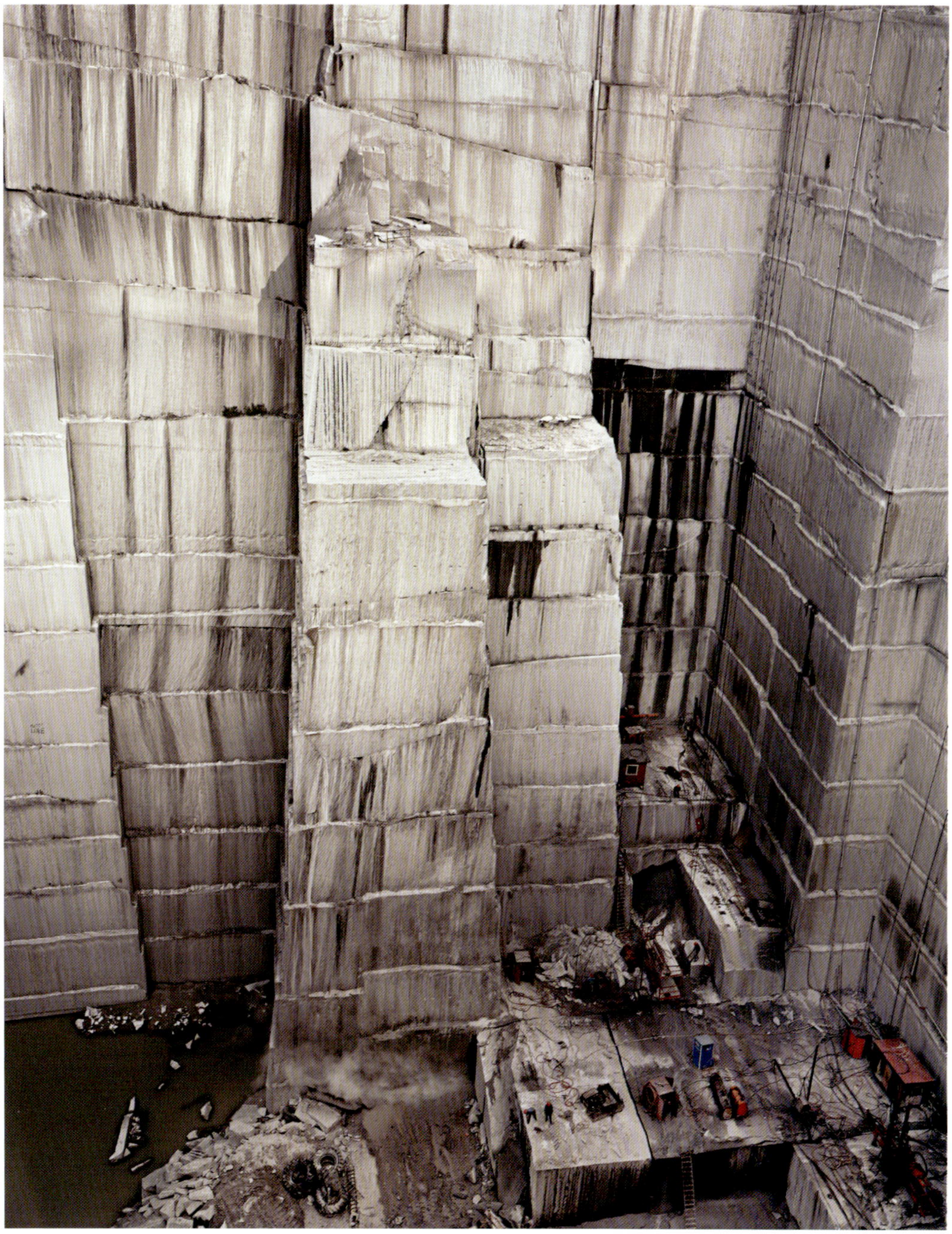

Plate 19
Edward Burtynsky
Rock of Ages #39, Active Section, E. L. Smith Quarry, Barre, Vermont, 1991
Digital chromogenic color print
Courtesy of the artist

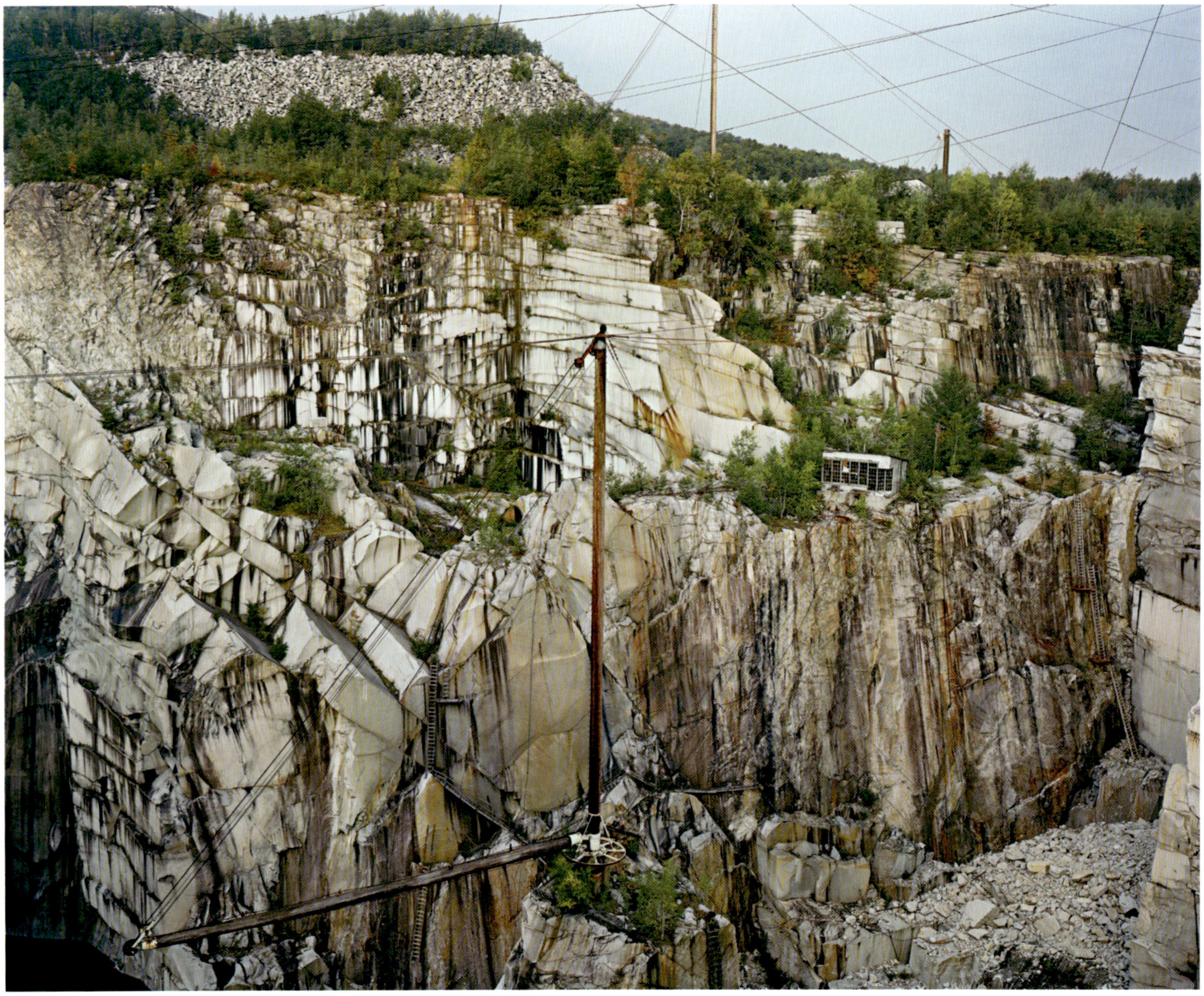

Plate 20
Edward Burtynsky
Rock of Ages #42, Abandoned Section, Wells-Lamson Quarry, Barre, Vermont, 1991
Digital chromogenic color print
Courtesy of the artist

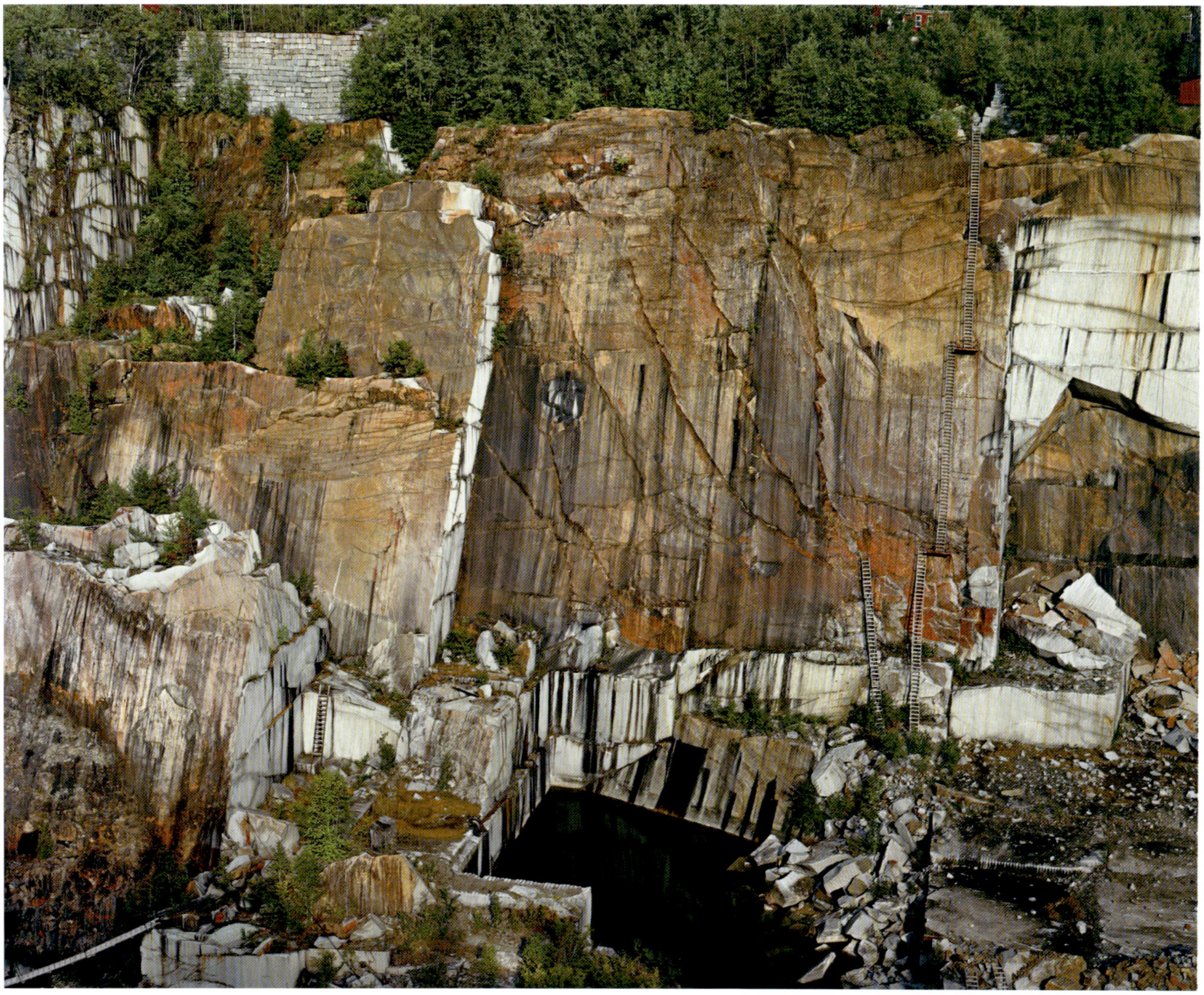

Plate 21
Edward Burtynsky
Rock of Ages #45, Abandoned Section, Rock of Ages Quarry, Barre, Vermont, 1991
Digital chromogenic color print
Courtesy of the artist

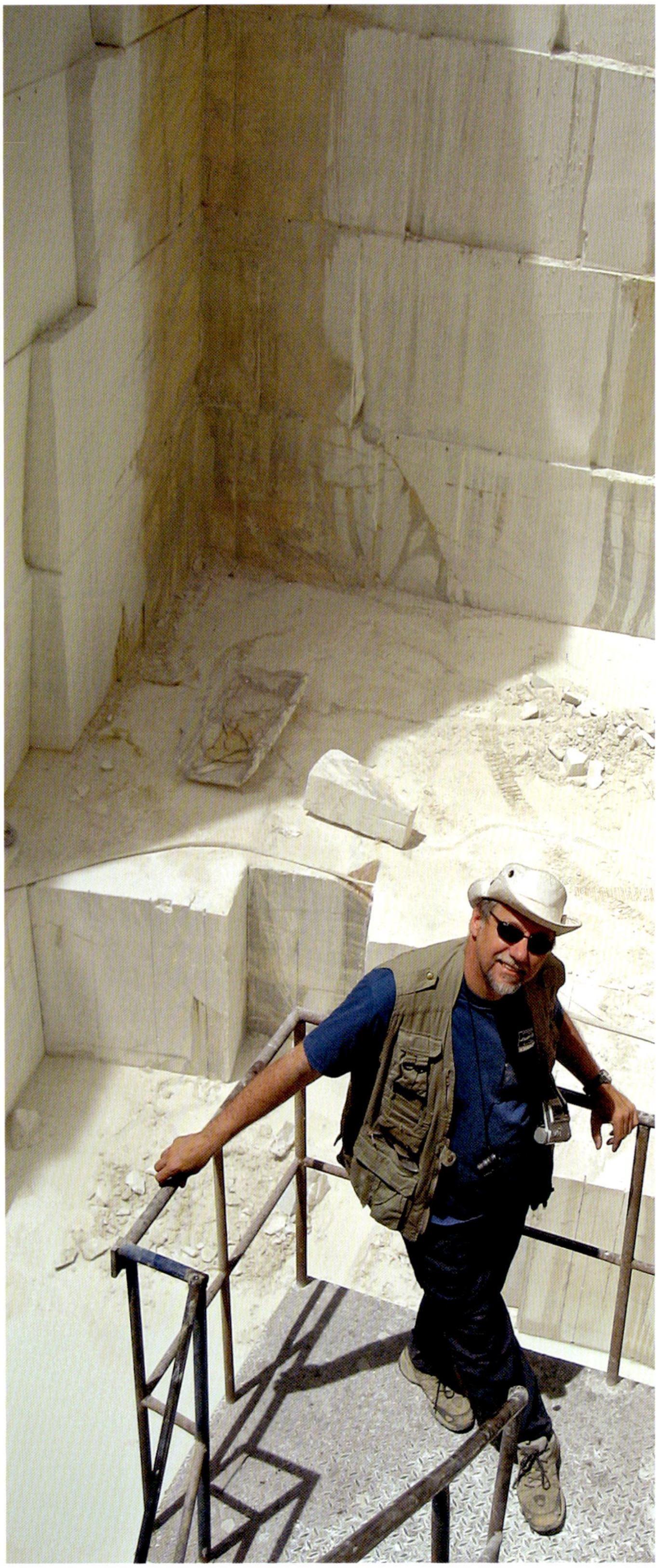

5
An Interview with Edward Burtynsky

Juliette Bianco

Juliette Bianco (JB): You've been taking photographs since you were eleven years old. Can you tell me about those early years—what was it about photography that drew you in?

Edward Burtynsky (EB): I was always interested in the arts, whether it was painting, sculpture, or building a go-cart. I found myself happiest when I was making things. When I was eleven, a widow whose husband had been a photography enthusiast was looking for a good home for all of his equipment, and I ended up with the contents of a complete darkroom. One box had three cameras in it, along with a bulk loader, which gave me the opportunity—which I think kids have today with digital cameras—to take many photographs of a single subject without the parents saying that you were wasting film.

The first pictures I ever took were of my dog "Tippy" jumping in the snow. I was fascinated to see that a couple of them were really great and the rest were lousy. That was interesting to think about—what made some pictures better than others? I intuited that the world appears very different in photographs than it does in reality. Although a photograph represents a captured moment in reality, each one is disassociated from that reality and subjected to other sets of criteria, which make it a better or a worse picture of the world. So I began to take a lot of pictures, just to see what the world looked like as an image. When I started to run out of the chemicals to process the film, I took pictures at the community center of all the kids in their events and got my sister to sell them to the parents. This became quite a good business when I was thirteen years old—I was able to underwrite my ability to take photographs, something I continue to do today.

JB: *Quarries* was your first project that was theme based. What is it about working with a theme—now a signature of your photography—that you find compelling?

EB: I always tend to work in series and on themes. Even before *Quarries*, I was looking at greenhouses, taxidermy workshops, railcuts, mines, and homesteads. What I liked about the mining series was that you could really see the impact of that industry—that void was one of the few places you could witness the scale of contemporary intervention in the landscape. I started to wonder about where else that sort of thing was occurring, and I thought of our cities, which are made from stone that is kept intact, whereas mining ore is not about maintaining the integrity of the stone. The type of excavation that resulted from dimensional stone seemed to indicate that there had to be a more orderly removal of the materials than at an ore mine. The idea I had was that I might be able to find the reverse of a skyscraper somewhere, an inverted pyramid where the blocks were being removed. So I pursued that.

Quarries therefore became the first series where I started purely with an idea. I tried to find quarries in Canada but they were not very developed. The Canadian quarrymen sent me to Vermont to see the

Edward Burtynsky at Makrana Marble Quarries, Makrana, India, November 2000. Photograph by Jim Panoa.

quarries—both active and abandoned—in Barre, Rutland, and Rochester. I was just as interested, if not more so, in the abandoned quarries as I was the active ones. The abandoned voids were bigger, but also there was the patina on the walls and nature springing back through cracks in the rocks. These offered a whole new range of visual possibilities as landscapes than did the more industrially active ones.

JB: When we met in February 2011, I was very impressed by your knowledge of the stone quarry industry and its techniques, which I gather you learned while working on this project twenty years ago. Why is it important to you to have that level of understanding of your subject?

EB: As a photographer of the industrialized world, where corporations and the collective needs of society are being addressed by these larger interventions into the landscape, the more I understand about that landscape, the better pictures I can make. Not only that, but the more I can talk about them and bring understanding and knowledge to the viewer about what they represent—why that stone, why in this place, why the different techniques—the better I am able to capture the moment with which the photograph intersects. As time goes by, this becomes more and more important. For example, if we look at a nineteenth-century street scene in New York City, no one at the time was thinking that they were leaving anything special for the future, because that's how everything looked then. But now they become references. I appreciate that photographs do carry that kind of cultural baggage with them through time. This is especially important when you are creating photographs that are mirrors of our world, which is what I do, instead of mirrors of our psychology, which is another way of using photography. Currently, I'm working on the theme of water, so I'm reading and studying everything possible about water, from scientific reports, to stories, to films. This deeper appreciation for the subject will help me make decisions—why make this picture and not that? Why go to this place instead of that? Knowing more about the subject provides an opportunity not only to make a better picture visually, but also to bring out layers of important information.

JB: This exhibition places your *Quarry* photographs in a historical context, telling the story, for example, of the Italian immigrant experience in Vermont. What are some of the virtues of revisiting your photographs in their social, and in this case geological, context?

EB: I learned that a lot of the Italians from the Carrara region came and even purchased a number of the quarries. In fact it was a quarry administrator in Vermont that said that I should go to Carrara to see where they had come from, which in 1993 became my first body of work outside of North America. So I do think the historical layer of information is interesting, and it is certainly legitimate. It doesn't directly speak to my photographs in the quarries, of course—I wasn't interested in the labor force or who owned the quarries; I was interested in the human need for stone, and where we go to get that stone, and it happened to be in these places. But the fact is that there's a whole other side to the story of who these people were, and it adds something that is not visible in my photographs.

JB: What is the relationship between the subject matter and the purely visual in your work—how do you strike that balance? Do you start with a story in your projects, and then find the photographs to tell it, or do you start with what interests you visually? Have you ever scuttled a project for having one but not the other?

EB: Photography really needs both. I'll often not go forward if the visuals are not working. If it just becomes "a document" of a place, then the photograph doesn't transcend its indexical or informational layer. I haven't positioned myself as a traditional documentary photographer. The photographs have to have a lot of things working correctly for me to stop and commit to spending a lot of time in one location. Sometimes I try and it doesn't work. That doesn't mean it was a waste of time. Five years later, I might go back and try shooting the same thing another way and I might find another combination of the idea, the content, and the form in equal measure to go forward.

JB: You came at your current work from an initial impulse to photograph the pastoral landscape. Many of your photographs, especially the earliest ones, were devoid of actual human beings. Your *China* series is full of people, but they are represented more as part of the landscape. What is the relationship between people and the landscape in your photography?

EB: What's different in the *China* series is that industrial work in China—the transplanting of the industrial revolution to a country that has in turn become the manufacturer for the world—was removed a step from the direct interventions in the landscape where the actual materials are being taken out at their primary source. In the manufacturing sector images, you are seeing an implied landscape—the materials still come from nature, but here they are being turned into products. I started this project because I saw the decline of manufacturing in Canada and other parts of North America and knew that it was heading over to China, but there weren't any telling images of that phenomenon. The idea of the image on the cover of the *China* book was that the human scale—eight thousand people lined up before work—was part of the new landscape (fig. 5.1). I wasn't in pursuit of the worker as part of a traditional photojournalistic story. I was interested in the uprooting of the industrial revolution and its manifestation in China at scale. It is not the individual human story but rather the collective human-impact story that I'm talking about.

JB: The stone quarries have become part of the accepted topography of Vermont, maybe because of their age, because they look more "natural," or because there are now so many more dangerous places on earth where extractions of other kinds are taking place, whether it is oil or heavy metals. Yet they are a violation of the landscape, too. Do you feel that the impact of your *Quarries* photographs is different now than it was twenty years ago?

EB: In the sense of human acts perpetrated upon nature, the quarries are quite frankly pretty low impact and high usability, in terms of what comes out of them. The quarried stone is immediately useful, and stone is a fairly benign extraction compared to gold or copper mining, where arsenic in the former and cyanide in the latter leaches into the landscape and leads to poisoning downstream. Once they are in the water table, these things affect land far beyond the parameters of the activity. Quarrying for stone, they use a little bit of power/fuel for the drill, and a bit of blasting to break out the blocks. When they are finished, the water stays (see plate 18), and the quarry fills up with it. Quarries often even become swimming holes. I remember swimming in a stone quarry as a kid. Other life starts to evolve around the quarry as well, and so they are not the worst thing we do in the landscape.

All living things take from the land—it's just that we are at the top of the food chain, helping ourselves to a lot of what's out there. It's not the fact that we take, but the scale of the taking that is starting to transform our planet.

JB: Congratulations on the exhibition of your new work, *Dryland Farming*, from Spain. I remember seeing you proofing some of those photographs in your studio in February 2011. What's the impact there?

EB: That series is quite exciting, and I just did another series on farms in the United States that I'm also very excited about. Farming is something that does seem very benign and is very important to us, but it has transformed the planet more than any other single human activity. Look how many forests have been lost, prairies have been lost, wetlands have been drained for it. A nice farm with a barn on it represents the single most destructive human impact on the planet.

JB: The pesticides washing into the rivers, lakes, and oceans are so dangerous.

EB: Not only the pesticides but the nitrates as well. The algae blooms at the mouth of the Mississippi have rendered an almost one-hundred mile radius around it dead as a result of high levels of nitrates in the water, which then promote algae growth, which then depletes the oxygen within the water, which

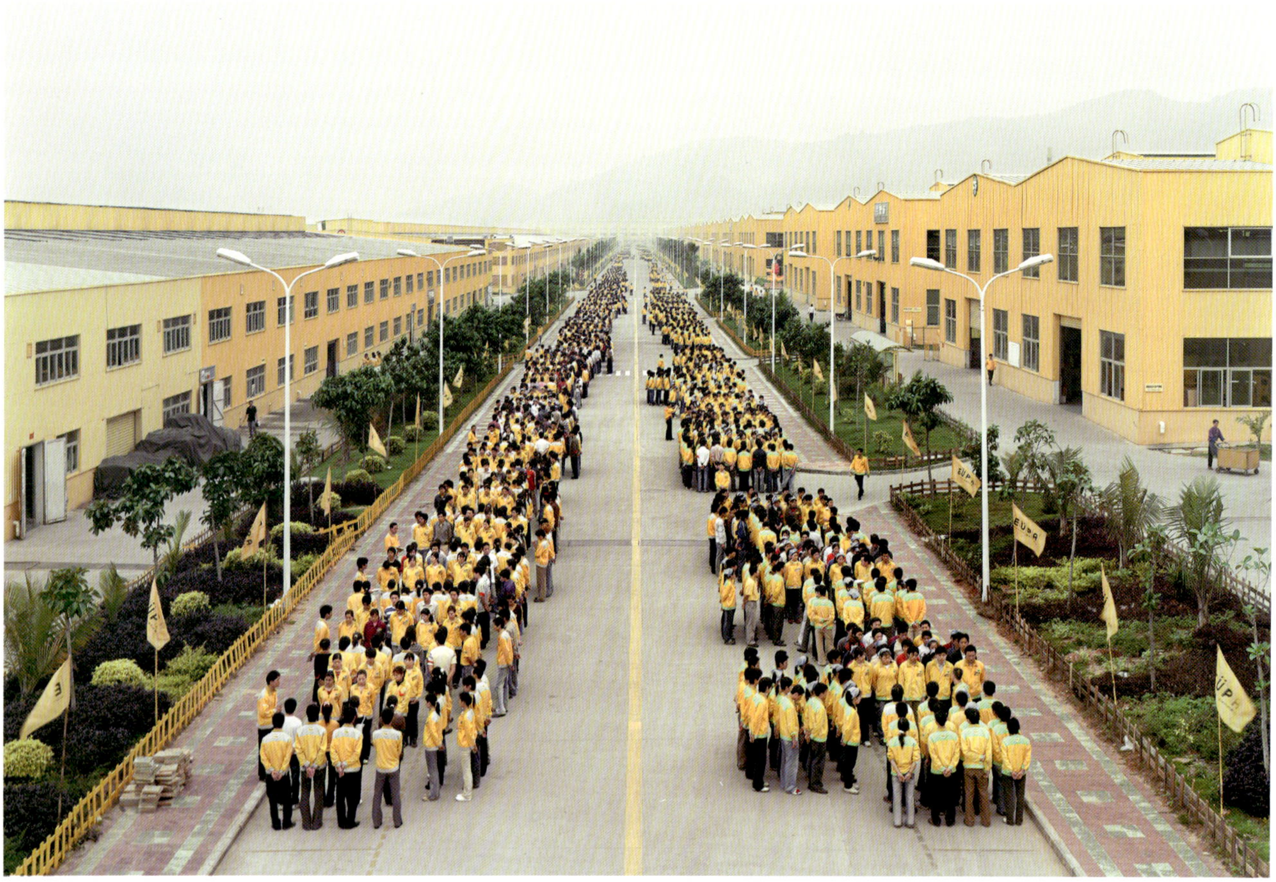

Figure 5.1
Edward Burtynsky
Manufacturing #18, Cankun Factory, Zhangzhou, Fujian Province, 2005, digital chromogenic color print. Courtesy of the artist.

then becomes anaerobic, and then the fish suffocate because there is no oxygen.

JB: In your *Quarries* photographs you distinctively put the horizon line very close to the top, if you use one at all. Can you talk about how you compose your images?

EB: I think in a classical sense I do look for a kind of cohesion, symmetry, or "unitedness" within the frame. I like framing that is tightly bound, whether it is deeper space or all-up-front space. But I arrive at each composition like a safecracker—you go to a new safe, and you don't know exactly what the numbers are, and you have to be very sensitive to hear those tumblers drop. Similarly, I go into a place because it's the right time of year, when I think the light is going to be optimal for the subject that I am trying to photograph. Is this a morning shot because the sun is at my back? I don't usually shoot backlit with lots of shadows, or in full sun. So those are the basic big blocks, and then the rest is about seeing the image.

Oftentimes it is even hard for me to see the potential of an image until I shut out my peripheral vision and hone down a composition that I think has the ability to function on a wall. Images in books or on the Internet are surrogate experiences of the large print on the wall, which is to me even more interesting than the experience of reality. The act of capturing the image to begin with is never as interesting as the resulting picture.

People think that taking these photographs must be very clear and exciting, but it's not. When I'm in that moment in that place waiting for all the conditions to align, I don't even know what I have until the light and all these other elements come together. It would not be unusual to have an eight-by-ten camera set up and then wait around for forty-five minutes to two hours, shooting perhaps a few times along the way. Then when I make the contact prints, it becomes obvious where it all came together in one frame.

JB: Do you stay connected to the places where you travel to photograph? Do you keep up with what is happening at those locations, or do you tend to move on by necessity or practice?

EB: I tend to stay connected to some places. I cover a lot of subjects and sometimes I end up staying connected with certain individuals, like the guy who helped me get into the Carrara quarries and acted as my translator, for example. I still call him up whenever I'm in Italy, and since then he's helped me with other projects there. Some people become instrumental to my ability to function in places like China and India or other places where I don't speak the language, or know the territory. Those are key relationships.

JB: In the film about your work, *Manufactured Landscapes*, you were at a coalfield in China and were told that you were not allowed to take pictures at the site. The company representative said that it was very gloomy anyway, and you would not get beautiful pictures. Your assistant said that you would make it beautiful. What is your relationship to the idea of beauty in your photographs?

EB: I tend not to use the word "beautiful." I'd be more inclined to use terminology like "visually compelling." The problem with the word *beauty* is that in the world of fine art, it is often understood to refer to something that is decorative, that no longer carries weight or content. The twentieth century in a way represents art trying to distance itself from beauty. I'm aware of that. In my work I am entering places that people traditionally would never feel that you could experience an aesthetic moment, a moment where reality transcends itself—but it happens.

I was always interested in the idea of a collective consciousness, in the Jungian sense—that there are archetypes and things we all share, some interesting, cosmic hardwiring throughout all our brains; that when we encounter a compelling image, there is a universality that draws us in. It's not culturally specific—meaning not just in the United States or Canada but also in Mexico, or Europe, Africa, China. I've been to most of these places with my work and people all have a similar experience. It truly is a universal experience—composition, design, color, et cetera all working together to make that happen. But beauty is culturally specific. The kind of images I make are more about ways in which we interpret the world, and

I think landscape is one of those few forms that all of us can experience in this way I described.

I shot tire piles, for example. No matter how you shoot a tire pile, it is hard to say that it's beautiful. It's still a pile of tires. So I don't think my photos speak to beauty as much as they do to some larger consciousness. That's where I want to focus my work, because photography is so broad and can be interpreted in so many ways. Language is something that shuts out so many people, whereas images open up to all people. That's why I struggle with postmodernism—it seemed to take art to a language that was so symbolic and rigorous that only a few people could understand it. If you didn't have the context of the artist's own thoughts, there was no way you were getting it. There are very few rarified people who can speak to what that's all about, but I want my art to transcend cultures and still have a narrative force and meaning, without being trite.

JB: You are drawn to subjects that could be construed as political, either because they show images of environmental degradation or because they depict people whose lives are compromised by industry. Yet you indicate that you don't have political intentions. What do you want people to think about when they look at your photographs?

EB: I'm not so naïve as to think that there is not political content in my work. There is a kind of ethical obligation I have, in that a lot of people who let me gain access did so with the understanding that I was making images that were not an indictment of what they were doing, but more of a way of telling a narrative. These images can be interpreted as the good, the bad, *and* the ugly. You can look at them in all kinds of ways—if you didn't have all those factories in China, there would be tens of millions of people mired in poverty. The inflation rate in North America would have continued to go up if we continued to manufacture here. So one way to look at it is that China has helped keep America's inflation rate down.

I can't control interpretation, but I can certainly control the message, and I strive to be fairly neutral about it. At the same time, I understand that all things have a political reading, and there could be industrial, technological, or human-use stories for any one of my subjects. There is a twentieth-century aesthetic layer to the work as well. For me to give the message a political spin as it goes out the door would be to condemn it to one way of being seen. I'm far more interested in the fact that the meaning can be more open to interpretation. The work speaks to more than the political underpinnings that have allowed those places that I photograph to exist. No matter what, to allow humans to get to a population of seven billion, a lot of things had to have happened. Politics is just one component of a much larger series of events.

JB: What do you think is the distinct role for photography as a medium?

EB: What I am trying to do with photography is to interpret, and raise consciousness about, our world in a more decisive way. The focus of a still image is a powerful means of visually dissecting a space, and it allows us to be analytical. For example, fashion uses photography all the time. Watching a fashion model walk down the runway in real life, or even on TV, is not the same as seeing that same clothing in a glossy magazine, where you can break it down, see the stitching, see how it is composed. You won't get that in real life, or in a motion picture. Photography allows us to stop, so our minds can digest, interpret, and then break apart imagery very differently.

JB: Do you have a "dream project" out there somewhere?

EB: There are always things that linger, and I am slowly checking them off. I don't give up easily. My feeling is that a good idea will not disappear. It will still feel like a good idea one or two years from now. It's like an icon in the dock on the bottom of your Mac screen, bouncing up and saying, "Look at me, there's something you need to attend to over here." Ideas are like that, they keep popping back into your head. Often enough when this happens, you have to do something about it and start the process of unpacking it, figuring out how to do it and where to go next.

Selected Bibliography

Compiled by Chanon Praepipatmongkol

Ballamingie, Patricia, Xiaobei Chen, Eric Henry, and Diana Nemiroff. "Edward Burtynsky's China Photographs—A Multidisciplinary Reading." *Environments Journal* 37, no. 2 (2009/2010): 67–92.

Blackwell, Lewis. *Photo-Wisdom: Master Photographers on Their Art.* Auckland: PQ Blackwell/Chronicle Books, 2009.

Burtynsky, Edward. *Australian Minescapes.* Welshpool: Western Australian Museum, 2009.

——. *Oil.* Essays by Paul Roth, Michael Mitchell, and William E. Rees. Ed. Marcus Schubert. Göttingen: Steidl, 2009.

——. *Quarries.* Essay by Michael Mitchell. Göttingen: Steidl, 2007.

——. *China.* Essays by Ted Fishman, Mark Kingwell, Marc Mayer, and the artist. Göttingen: Steidl, 2005.

——. Contributions by Gary-Michael Dault and Lucas Lackner. *Before the Flood.* Toronto: Self-published, 2003.

——. *Edward Burtynsky: Works from Two Decades, 1981–2001.* Kitchener: Kitchener-Waterloo Art Gallery, 2002.

——. *Breaking Ground.* Canadian Museum of Contemporary Photography, 1988.

——. "Five Views of Catastrophe." *Queen's Quarterly* 117, no. 4 (winter 2010): 573–83.

Callahan, Sean. "Edward Burtynsky: Multiple Viewpoints." *Smithsonian* 33, no. 1 (April 2002): 93–97.

Dault, Gary M. "The Eleventh Hour in Photography." *Canadian Art* 20, no. 1 (spring 2003): 56–62.

Diehl, Carol. "The Toxic Sublime." *Art in America* 94, no. 2 (February 2006): 118–23.

Forum for Urban Design. *Shanghai: Explosive Growth.* New York: Forum for Urban Design, 2002.

Freedman Gallery. *Industry and Entropy: Edward Burtynsky, Mark Ruwedel, Robert Smithson.* Reading: Freedman Gallery, Albright College, 2003.

Jack, Ian. "Edward Burtynsky—The Evidence of Man." In *This Overheating World.* New York: Granta, 2003.

John, Knechtel, ed. *Fuel.* Cambridge, Mass.: MIT Press, 2008.

Lala, Kisa. "Interview with Edward Burtynsky: Photographing a Planet on the Edge." Spread ArtCulture blog, September 30, 2010. Available at www.spreadartculture.com/2010/09/30/interview-with-edward-burtynsky.

Natural World Museum and United Nations Environment Program. *Art in Action: Nature, Creativity and Our Collective Future.* San Rafael: Earth Aware Editions, 2007.

Paton, Todd. *The Rock of Ages Story.* Barre: The Rock of Ages Corporation, 2003.

Pauli, Lori. *Manufactured Landscapes: The Photographs of Edward Burtynsky.* Essays by Mark Haworth-Booth and Kenneth Baker. Ottawa: National Gallery of Canada, in association with Yale University Press, 2003.

Sasse, Julie, Emily Handlin, and Lindsay Russell. *Trouble in Paradise: Examining Discord Between Nature and Society.* Tucson: Tucson Museum of Art, 2009.

Sultan, Terrie, Nick Flynn, David Pagel, Colin Gardner, and Claudia Schmuckli. *Damaged Romanticism: A Mirror of Modern Emotion.* Houston: Blaffer Gallery, 2008.

Torosian, Michael. *Residual Landscapes: Studies of Industrial Transfiguration.* Toronto: Lumiere Press, 2001.

Wolfe, Ann M., ed. In *The Altered Landscape: Photographs of a Changing Environment.* New York: Skira Rizzoli Publications, 2011.

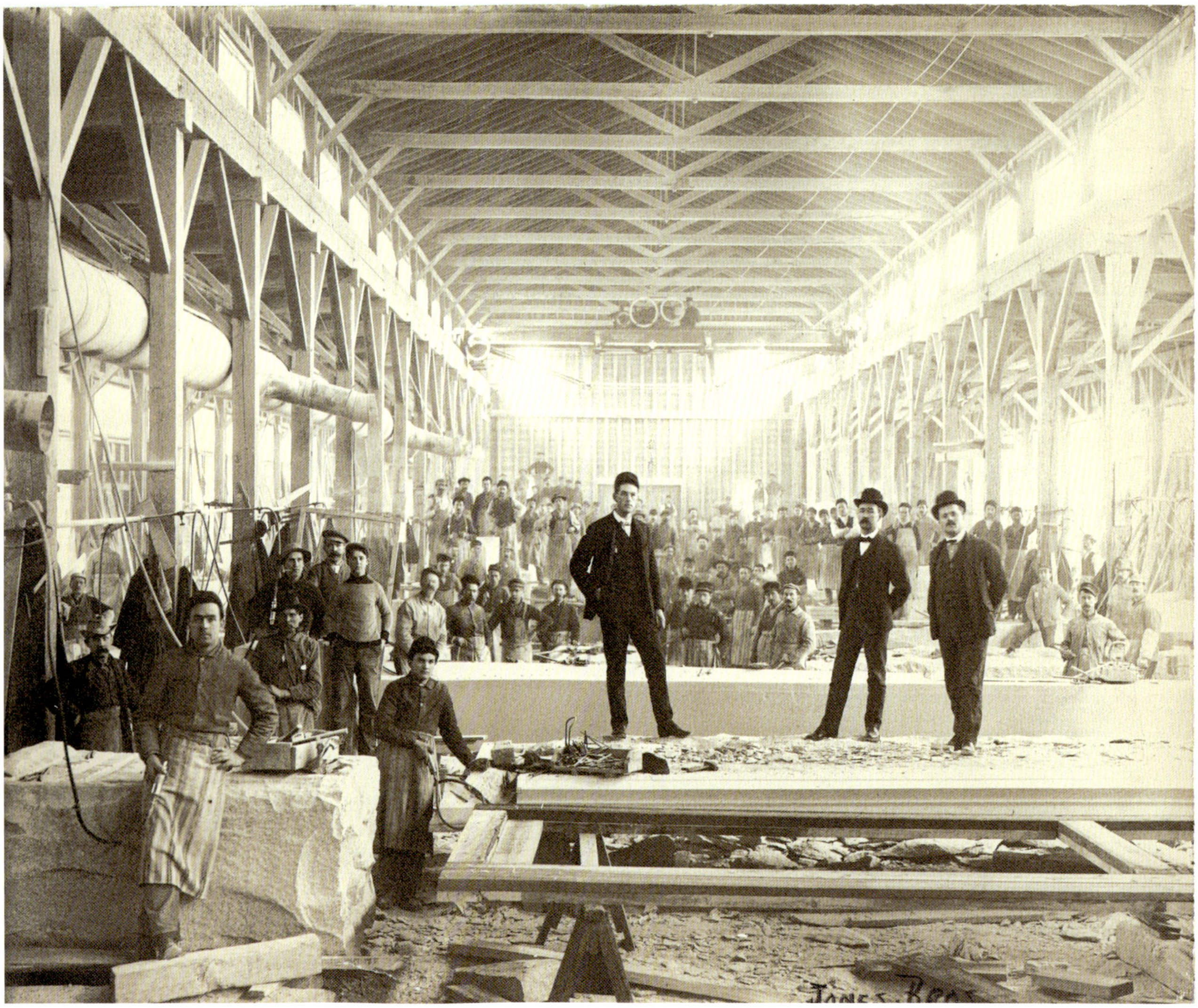

Photographer unknown
Jones Brothers Plant
About 1900
Vintage photograph
Courtesy of the Aldrich Public Library, Barre, Vermont

Exhibition Checklist

1 Edward Burtynsky
Marble Quarries #001, Rutland, Vermont 1991
Chromogenic color print
Sheet: 28 x 35 inches
Image: 25 x 32 inches
Lent by Dennis Sughrue
(plate 1, page 18)

2 Edward Burtynsky
Abandoned Marble Quarry #18, Rutland, Vermont, 1991
Digital chromogenic color print
48 x 64 inches
Courtesy of the artist
(plate 2, page 19)

3 Edward Burtynsky
OMYA #51, Calcium Carbonate Waste Area, Proctor, Vermont, 1991
Digital chromogenic color print
48 x 64 inches
Courtesy of the artist
(plate 3, page 20)

4 Edward Burtynsky
Vermont Marble Company #52, Vermont Marble Company, Rochester, Vermont, 1991
Digital chromogenic color print
48 x 64 inches
Courtesy of the artist
(plate 4, page 21)

5 Edward Burtynsky
Danby Marble Quarry #1, Underground Quarry, Danby, Vermont, 1995
Digital chromogenic color print
39 x 52 inches
Courtesy of the artist
(plate 5, page 22)

6 Edward Burtynsky
Danby Marble Quarry #2, Underground Quarry, Danby, Vermont, 1995
Digital chromogenic color print
39 x 52 inches
Courtesy of the artist
(plate 6, page 23)

7 Edward Burtynsky
Carrara Marble Quarries #20, Carrara, Italy, 1993
Chromogenic color print
Sheet: 29 5/8 x 36 5/8 inches
Image: 26 3/4 x 33 3/4 inches
Lent by Raphael and Jane Bernstein Collection
(plate 7, page 36)

8 Edward Burtynsky
Carrara Marble Quarries #26, Carrara, Italy, 1993
Chromogenic color print
Sheet: 36 5/8 x 29 5/8 inches
Image: 33 3/4 x 26 3/4 inches
Lent by Raphael and Jane Bernstein Collection
(plate 8, page 37)

9 Edward Burtynsky
Carrara Marble Quarries #24, Carrara, Italy, 1993
Chromogenic color print
Sheet: 28 x 35 inches
Image: 25 x 32 inches
Lent by Daniel Bernstein and Claire Foerster
(plate 9, page 38)

10 Edward Burtynsky
Carrara Marble Quarries #25, Carrara, Italy, 1993
Chromogenic color print
Sheet: 28 x 35 inches
Image: 25 x 32 inches
Lent by Daniel Bernstein and Claire Foerster
(plate 10, page 39)

11 Edward Burtynsky
Rock of Ages #1, Active Section, E. L. Smith Quarry, Barre, Vermont, 1991
Digital chromogenic color print
48 x 64 inches
Courtesy of the artist
(plate 11, page 49)

12 Edward Burtynsky
Rock of Ages #4, Abandoned Section, Adam-Pirie Quarry, Barre, Vermont, 1991
Digital chromogenic color print
64 x 48 inches
Courtesy of the artist
(plate 12, page 50)

13 Edward Burtynsky
Rock of Ages #6, Abandoned Section, Rock of Ages Quarry, Barre, Vermont, 1991
Digital chromogenic color print
48 x 64 inches
Courtesy of the artist
(plate 13, page 51)

14 Edward Burtynsky
Rock of Ages #7, Active Granite Section, E. L. Smith Quarry, Barre, Vermont, 1991
Digital chromogenic color print
48 x 64 inches
Courtesy of the artist
(plate 14, page 52)

15 Edward Burtynsky
Rock of Ages #12, Abandoned Section, Adam-Pirie Quarry, Barre, Vermont, 1991
Digital chromogenic color print
48 x 64 inches
Courtesy of the artist
(plate 15, page 53)

16 Edward Burtynsky
Rock of Ages #17, Abandoned Section, Adam-Pirie Quarry, Barre, Vermont, 1991
Digital chromogenic color print
48 x 64 inches
Courtesy of the artist
(plate 16, page 54)

17 Edward Burtynsky
Rock of Ages #19, Granite Section, Rock of Ages Quarry, Vermont, 1991
Chromogenic color print
40 x 50 in.
Lent by the Middlebury College Museum of Art, Vermont: Purchase with funds provided by the Walter Cerf Art Fund; 2008.034
(plate 17, page 55)

18 Edward Burtynsky
Rock of Ages #25, Abandoned Section, Adam-Pirie Quarry, Barre, Vermont, 1991
Digital chromogenic color print
48 x 64 inches
Courtesy of the artist
(plate 18, page 56)

19 Edward Burtynsky
Rock of Ages #39, Active Section, E. L. Smith Quarry, Barre, Vermont, 1991
Digital chromogenic color print
64 x 48 inches
Courtesy of the artist
(plate 19, page 57)

20 Edward Burtynsky
Rock of Ages #42, Abandoned Section, Wells-Lamson Quarry, Barre, Vermont, 1991
Digital chromogenic color print
48 x 64 inches
Courtesy of the artist
(plate 20, page 58)

21 Edward Burtynsky
Rock of Ages #45, Abandoned Section, Rock of Ages Quarry, Barre, Vermont, 1991
Digital chromogenic color print
48 x 64 inches
Courtesy of the artist
(plate 21, page 59)

22 Photographer unknown
Wells-Lamson Quarry
About 1908
Vintage photograph
Courtesy of the Aldrich Public Library, Barre, Vermont
(fig. 1.1, page 2)

23 Photographer unknown
Jones Brothers Plant
About 1900
Vintage photograph
Courtesy of the Aldrich Public Library, Barre, Vermont
(page 68)

24 Photographic Studio of F. W. Sherburne, Barre, Vermont
In Barre; part of Stanford Mausoleum; weight of stone 54 tons
1888
Vintage photograph
Courtesy of the Aldrich Public Library, Barre, Vermont
(figure 3.2, page 27)

25 Photographic Studio of F. W. Sherburne, Barre, Vermont
Barre Railroad
About 1890
Vintage photograph
Courtesy of the Aldrich Public Library, Barre, Vermont
(figure 3.1, page 27)

26 Photographer unknown
Granite quarry and workers, Consolidated Granite Co., Eagan Quarry, Barre, Vermont
Date unknown
Photographic copy from glass slide negative
Courtesy of the Aldrich Public Library, Barre, Vermont
(figure 3.3, page 30)

27 Photographer unknown
Granite quarrier using pneumatic drill
About 1900
Photographic copy from negative
Courtesy of the Aldrich Public Library, Barre, Vermont
(figure 3.4, page 30)

28 Photographer unknown
Veronese band
1916
Photographic copy from negative
Courtesy of the Aldrich Public Library, Barre, Vermont
(figure 3.7, page 32)

29 Photographer unknown
Il Fornaretto di Venezia (The Baker of Venice) cast from the production by La Vecchia Filodramatica Society at the Barre Opera House
1902
Photographic copy from negative
Courtesy of the Aldrich Public Library, Barre, Vermont
(figure 3.8, page 32)

30 Photographer unknown
Picnic in a field above the North End in Barre, Vermont
Date unknown
Photographic copy from negative
Courtesy of the Aldrich Public Library, Barre, Vermont
(figure 3.6, page 31)

31 Photographer unknown
Il Primo Maggio (The first of March—Italian Labor Day), Barre, Vermont
1906
Photographic copy from negative
Courtesy of the Aldrich Public Library, Barre, Vermont
(figure 3.11, page 33)

32 Photographer unknown
Interior of the Corti and Novelli shed
(Corti is at the far left; Novelli is at the right, seated)
About 1902
Photographic copy from negative
Courtesy of the Aldrich Public Library, Barre, Vermont
(figure 3.5, page 30)

33 Photographer unknown
Elia Corti monument, Hope Cemetery, Barre, Vermont, carved by William Corti and John Comi
About 1903
Photographic copy from negative
Courtesy of the Aldrich Public Library, Barre, Vermont
(figure 3.12, page 34)